Spiritualities, ethics, and implications of human enhancement and artificial intelligence

Edited by
Christopher Hrynkow
University of Saskatchewan

Series in Philosophy
VERNON PRESS

www.vernonpress.com

In the Americas:
Vernon Press
1000 N West Street,
Suite 1200, Wilmington,
Delaware 19801
United States

In the rest of the world:
Vernon Press
C/Sancti Espiritu 17,
Malaga, 29006
Spain

Series in Philosophy

Library of Congress Control Number: 2019947991

ISBN: 978-1-62273-823-6

Irene J. Dabrowski and Anthony L. Haynor (authors of Chapter 3) wish to thank Ignatius Press for permission to quote extensively from Henri de Lubac's *The drama of atheist humanism* (pages 400, 401, 403-409, 411, 414, 417, 420-422, 451-452, 468). To order this book, go to https://www.ignatius.com/The Drama-of-Atheist-Humanism-P528.aspx.

Extended quotations from Pope Francis' *Laudato Si'* presented in Chapter 8 used with official permission (N. 0003803/ 2019). © Libreria Editrice Vaticana, 2015.

Cover design by Vernon Press using elements designed by Freepik.

Table of contents

Setting the stage for conversations about human enhancement, artificial intelligence, and spirituality

1.

Engaging issues at the intersection of human enhancement, artificial intelligence, and spirituality

Christopher Hrynkow

University of Saskatchewan

The majority of the papers in this volume have their origins in the inaugural conference of the Saskatchewan Center for Science and Religion, held in Vancouver BC in November 2017. This gathering was generously supported by the center's founding benefactor Wei Lei. The goal of the conference was to initiate and deepen interdisciplinary conversations about human enhancement, artificial intelligence, and spirituality. Those able to attend were treated to papers, given by academics working from diverse perspectives including scientists and scholars of religion, who considered the spiritual implications of issues surrounding human enhancement and artificial intelligence. A wide-ranging dialogue ensued concerning topics like how artificial intelligence may augment spirituality or make it obsolete, the challenge of transhumanism for world religions, the religious-ethical implications of artificial intelligence, faith traditions' reactions to efforts to engineer human enhancement, and the opportunities and significant tensions emerging for religious traditions in light of recent developments in artificial intelligence research.

The conference opened with a memorable keynote address from Ray Kurzweil, a transcribed and edited excerpt of which performs the same function for this volume. Herein, Kurzweil situates his contribution in the area commonly named as transhumanism with references to religiosity and spirituality that will be of great interest to scholars in the field of religion and science. The interweaving of elements of his life story and family history with his views on the transformative potential of various information and intelligence technologies is fascinating. The keynote elicited some interesting questions for the audience and poignant responses from Kurzweil, a selection of which are also presented in edited form for the readers' consideration.

The subsequent chapters engage and contribute to the academic discourse surrounding human enhancement, artificial intelligence, and spirituality with foci on the authors' respective areas of expertise. These body chapters begin with a sharp contrast to the perspective represented in Kurzweil's contribution. Irene Dabrowski and Anthony Haynor compare their articulation of a Catholic worldview with what they illustrate, building upon the work of the Jesuit thinker Henri de Lubac, as atheist humanism's foothold within transhumanism. Their chapter is helpful in setting the stage for the reader to ponder ethical undercurrents circulating among transhumanists. It also names an alternative bound together with their articulation of a Catholic worldview. Next, in this section, Mark Graves calls our attention to some of the ethical issues that come into play when AIs interact with models of human morality and spirituality. Graves shares a particular concern with how these models offer a space for AIs to undertake tasks in support of human flourishing, while navigating the thorny area of suffering. He also poignantly identifies some of the tensions that accompany efforts to construct models of shared human values that could be employed by AIs who will have a different embodiment than *Homo Sapiens*. These models can, in turn, allow them to interact morally with humans. Graves raises the possibilities of a modeling or morality that can integrate shared memories and hope along with social, religious, and shifting interpretative elements building upon the contributions of an already existing community of interpretation concerned with AI spirituality. Employing cutting edge methodologies for analyzing moral content, he draws our attention to the possibilities of AIs ability to process texts quickly as a potentially fruitful path to establishing such models, a path he frames as analogous to formal human study of morality and religiously significant texts.

The present volume then turns to consider the implications of human enhancements within contemporary societies. It begins with Una Stroda employing Apple's well-known Siri aid as an entry point to consider the place of humor in a transhumanist future. The body of her chapter considers how laughter is treated in diverse faith and cultural traditions through a comparative lens. Stroda then intriguingly considers how select aspirations of transhumanists will shift laughter and its manifestations as we know them. This section of the book ends with Tracy Trothen's fine piece addressing the intersections amongst spirituality, religion, sport, and human enhancements. She presents several moral dilemmas in order to bring some of the latest developments and issues in human enhancement into dialogue with the theological concept of hope. Moreover, she relates both areas to the experiential elements of sport, from (1) performances that appear 'perfect' to external viewers to (2) the emic achievement of 'flow'. Trothen leaves her reader contemplating the importance of hope for framing and shifting debates concerning the ethics of human enhancements in sport.

Michael Caligiuri opens the section on technology and the moral human body with his impressive survey of Roman Catholic perspectives on the philosophical and ethical consequences of cyborg technology as they extend outwards from both lay and clerical bodies. In the process, he skillfully weaves together both secular and religious antecedents to such discussions in a manner that is sure to peak reader interest. Caligiuri concludes by articulating six principles for guiding human enhancement, which he positions as congruent with Roman Catholic moral teaching and thought. Next, Jacob Boss turns to look at the situation of what he informatively names as the '*petite cyborg*', who is distinguished by a fusing of the human with cybernetics that remains dependent on the global energy and capitalist systems. With reference to the diverse discourses drawn from sources including science fiction and the thought of Bruno Latour, he considers some possible futures that can arise from the horrors and possibilities of cybernetics. This section closes with a chapter typed by myself that employs the thought of Pope Francis on integral ecology and articulates its relationship to scientific endeavor. The results of that conversation are applied as a lens through which to consider both the promises and perils of radical life extension. A particular concern of this chapter is to place these promises and perils in relation to socio-ecological flourishing understood as flowing from humanity's status as embodied creatures existing in intertwined social and ecological contexts within an Earth community.

The section on worldviews and artificial intelligence begins with Braden Molhoek's exploration of the application of AI ethics to transhumanism with reference to Michael Waltzer's reflections on citizenship. Molhoek asks a number of poignant questions that set the stage for the insights presented in subsequent chapters. Molhoek's contribution concludes with a compelling treatment of some ways that religious engagement will be shifted by plausible developments in AI and transhumanism. Christopher Benek continues the focus on AI, looking at its effects on both technologists being drawn towards considering religiosity and theologians being drawn to reconsider humanity's role in the cosmos. Benek undertakes this exploration through his own three-part theological lens, leading to an argument in support of his participatory model. In this manner, he presents a theological hermeneutic aimed at allowing people to better themselves and their anthropological location in light of exponential developments in AI. This section ends with Peter Robinson unfolding his innovative research and some of its spiritual implications. Particularly noteworthy here is how Robinson challenges some of the misperceptions that people have about the ability and potential of computers to empathize. Further, he makes some intriguing connections to the concept of *imago dei* in his chapter.

In the penultimate chapter, Philip Butler explores concepts of spiritual enhancement in an intriguing way that focuses upon the ethical implications of spiritually augmenting brain-computer interface's ability to mimic the biochemical processes stimulated by psychedelics. The reader is sure be impressed by his in-depth knowledge of these areas and his ability to think through a fascinating set of ethical implications of this emergent transhumanist technology. This volume's final chapter examines the potential for harm, collateral risks, and some of the unintended negative consequences of intrusions into people's mental domains associated with the emerging field of neuro-technology. Alan Weissenbacher shows that these relatively unproblematized interventions hold coercive properties. In response, he argues for the importance of his criterion of 'acceptability across ideologies' to avoid undue manipulation of neuro-technology in the service of ideologies in their religious, consumerist, and political forms.

This introduction has given a flavor of what is on offer in this volume. Careful attention to each chapter will repay the reader with interest in the complex and pressing issues that come into play when considering the spiritual and religious implications of human enhancement and artificial intelligence. I sincerely hope that each reader's learning and reflection will be enriched in a comparable way to be my own experience in editing this volume. In this light, I invite you to enjoy the diverse, innovative, and intriguing voices that are brought to the fore in the pages that follow.

-Christopher Hrynkow, Toronto, June 2019

2.
Breaking the shackles of our genetic legacy

Ray Kurzweil

Inventor, Author, Futurist

Editor's Note: The following is an edited excerpt of Ray Kurzweil's captivating keynote speech given at the Saskatchewan Center for Religion and Science's inaugural conference on Spiritualties of human enhancement and artificial intelligence. That talk was peppered with cogent stories and relevant insights on issues at the intersection of religion, science, and spirituality. As such, the editor is grateful to Mr. Kurzweil for reviewing this written excerpt of his keynote and permitting its inclusion in the present volume. Although different in tone and approach than the rest of the contributions, it is hoped that academics and other interested parties will find the nuggets about religion and spirituality as they intersect with Mr. Kurzweil life and work intriguing and valuable.

Introduction

It is a pleasure to be with you. This is a very unique conference, and I appreciate being invited. I have thought about the intersection of spirituality and religion on the one hand and science on the other. Today, I will share some of my resultant ideas with you. To start, let me give you a little bit about my background. I will also try to share some of the key points of my thesis concerning evolution of technology and its intersection with humanity, while reflecting on the goals of this conference.

To tell my story properly, I need to go back to 1868. As you know, I am interested in radical life extension, but I am not quite that old. My mother's mother's mother, in 1868, started the first school in Europe that provided higher education for girls. It was a school in Vienna, called the *Stern Schule.* This was an important development. If you were a girl in mid-nineteenth-century Europe and were able to get an education at all you went through at most to the ninth grade. In contrast, the *Stern Schule* offered educational opportunities from kindergarten to 14th grade, through high school to 1st year in the college. It was very controversial. Many people asked, "What is the point of doing that?" It was a frequent challenge. My great grandmother went around Europe, lecturing on the importance of educating girls and women, and explaining the best approaches to that task. She became a very influential expert on the education of women in Europe. Her daughter, my grandmother, became an example of her philosophy and was the first woman in Europe to get a PhD in chemistry. My grandmother also lectured about both chemistry and the importance of educating women. She eventually took over the school. Between the two women they directed the *Stern Schule* for seventy years from 1868 to 1938. They then fled from the Nazi regime, after Hitler had annexed Austria in March of 1938.

They came to America with my mother, and in 1948, the year I was born, my grandmother wrote a book called *One Life is not Enough*, which does presage my interest in life extension. The book was an autobiography in which she talked about her histories and the history of the school. My grandmother wrote it out on a manual typewriter. When I was five, she showed me the book and the typewriter. As a five-year-old, I was more interested in the typewriter than in the book. I did take an interest in the book some years later. But to me, this typewriter was a magical machine. It could actually take a blank piece of paper and turn it into something that it looked like it came from a book. I had never seen anything like that—it was quite magical.

I was familiar with some parlor magic tricks, and would even put on a little magic shows. However, I realized that in the case of parlor magic if someone sees how the magic trick works, the magic goes away. But this was not the case with technology. I did not have the words to describe this phenomenon when I was five. Nonetheless, I figured out how this machine worked by following all the mechanical linkages—and it was still magical.

This inspired me to go through the neighborhood to gather little mechanical linkages. This was an era in which you would allow a five-year-old to go to through neighborhood and do this. Now you would probably get arrested if you allowed your five-year-old child to do that. I would bring back broken bicycles or electrical things such as broken radios and other little mechanisms that were discarded. In this way, I amassed a big inventory of little things, little gadgets and mechanisms. Then I had this idea: if I can just figure out how to put them together, then I could solve any problem. I remembered talking to these much older girls, I think they were ten, and explaining that I would be able to go to the moon and be able to solve all disease if I could just figure out how to put these things together. In response, they said, "You have quite an imagination, kid." And so, my inventions did not get much traction for a while. However, when I was eight, my first success came when I created a virtual reality puppet theater. I had a mechanical control station. I could control the virtual world from this centralized station with mechanical linkages. I could move characters on and off the stage, and move their arms, move the sun and stars. This was a very big success in the eyes of my fourth-grade classmates. When I was twelve, I discovered the computer. Hardly remarkable today, when you can walk through an airport and everybody over the age of two has a computer. But when I was twelve, there were only twelve computers in all of New York City. I was allowed to use the midnight to 8:00 A.M. shift at Flower-Fifth Avenue Hospital in Spanish Harlem, New York City. I would take the train down there at the age of twelve and program that computer at night. I later became the software engineer for the Head Start program, doing analysis to prove that preschool education was a good thing.

Let me mention a few more explicit connections between (1) all of this material and (2) the goal and theme of this conference. I have written a fair amount about the connections among technology, science, and spirituality. I will mention three ideas. My 1999 book was titled *The Age of Spiritual Machines.* So what did I mean by that title? Well the ultimate spiritual quality, and this is, I think, respected by everyone whether you formally believe in God or religion or not, is consciousness. So a conscious being is really the seed of value, namely, our actions are only important in so far they affect the conscious experience of conscious beings. It follows that non-conscious entities, no matter how interesting, are only important in so far as they affect the conscious experience of conscious beings. So who and what is conscious? That is a debate. We have accepted that humans are conscious, or at least some humans are. Not all humans seem to act conscious. The whole debate about animal rights is centered on whether they are conscious. I think my cats are conscious. Not everybody believes that. I think they have not met my cats. We will also have a debate about artificial intelligence and consciousness. AI does not yet have all the subtle cues that we associate with conscious behavior. However, the essence of my prediction is that by 2029, they will display these characteristics of consciousness. But are they conscious? I predict that human consensus will be that AIs [artificial intelligences] that are conscious. If we say they are not conscious, they will get mad at us, and they will be very smart so, we will not want that to happen. The essence of my prediction is not a scientific prediction; it is a philosophical one: I predict that these AIs in the 2030s, which will be at human level, will actually be conscious beings. This is one spiritual implication because the ultimate spiritual value is consciousness.

The second implication is that evolution is a spiritual process. What happens in both biological evolution and technological evolution? I have a graph in *The Singularity is Near* that shows the very smooth transition from biological evolution to technological evolution, in a continual acceleration of progress. Through evolution, entities get more knowledgeable; they get more intelligent and more creative. As we increased our neocortex, we became more capable of expressing emotion and loving sentiments; we become more loving. Well, how has God been described? I grew up in a Unitarian Church, where we studied all the different religions, and the theme was many paths to the truth. We noticed that even though they use different metaphors in different stories, the religions all described God in the same way as being unlimited using language like: God is all-knowing; God is infinitely creative, infinitely loving, and infinitely intelligent. What happens to these entities in the evolution is they become exponentially more capable in those very qualities, never achieving infinite levels, so they never become God. However, they exponentially become more God-like, and closer to God through this evolutionary process. Evolution speeds up. We are now at a very fast phase through technological evolution.

The third implication is the word infinity, which was regarded as a synonym for God. How many whole numbers are there? One, two, three, four... There is an unlimited amount, or an infinite number. No matter what whole number you describe, I could add one or two and get another one. So this goes on without a limit, and we say there are an infinite number of whole numbers. In the nineteen century, George Cantor, who was a devout Lutheran, got this theory and actually proved some theorems that there is more than one infinity; in fact, there is an infinite number of infinities. Cantor reported that he actually got this directly from God. For example, the number of real numbers, numbers like this the squares root of two, is greater than the number of whole numbers. Cantor argued that there is an infinite number of infinities and offered mathematical proofs for his conclusion. This was considered blasphemy by George Cantor's Lutheran theologian friends, because they really regarded infinity as a synonym for God. The idea that there could be more than one infinity means there is more than one God. That conclusion violates the tenet of monotheism. I have actually been thinking about that. I just wrote an essay, which is going to be my next book, *The Singularity is Nearer*. In that essay, I propose that there is actually an infinity, which I am defining mathematically, and showing that it is greater than all these other infinities, nothing transcends it; there is no infinity that is greater than it. I call it G. The Hebrew character aleph that George Cantor used to distinguish among different infinities is actually the first letter in the word angel. I conclude that there are these finite numbers which represent the finite beings. There is an unlimited number or an infinite number of infinite angels. There is one G that transcends all these other entities, and nothing transcends it. So I think George Cantor's Christian theologian friends would have been satisfied with that formulation. On that note, I'd like to take your questions. Thank you.

Question period

Question 1: You suggest that the Law of Accelerating Returns seems to be unlimited, citing similarities with Moore's Law and even evolution. Yet, there is constraint even with evolution since we have finite resources and with Moore's Law, some argue that as we get closer to the atomic level Moore's Law might actually be 'capped', therefore not unlimited. Are there examples within the Law of Accelerating Returns that as it meets natural resources (time, space, energy) that it also becomes limited?

Ray Kurzweil: I address this question in more detail in both *The Singularity is Near* (2009) and *How to Create a Mind* (2012). First, Moore's Law is not synonymous with exponential growth. As I mentioned, it is the fifth paradigm and not the first, nor is it the last to bring exponential growth to electronics and computing. We have embarked on the sixth paradigm, which is computing in three dimensions—Moore's Law has already slowed down, but the Law of Accelerating Returns has not. The human brain uses a very cumbersome computation mechanism that is a million times slower than electronics, but it gets its power from being organized in three dimensions, so we are now going into the third dimension with our circuits.

So yes, there are "limited" resources, but the minimum resources required to either compute a bit, remember a bit, or transmit a bit are vanishingly small and can be done with atom or less. But even using the limits that we have demonstrated, we can create circuits that are trillion times more powerful than the human brain. Thus, it is not as if we have a fixed amount of resources. Part of the progress that is behind these curves is that we use less and less resources. While we are not actually sure what the limit is, the limits we have demonstrated suggest we can go many decades to a point where we have vastly exceed human thinking.

Question 2: You seem optimistic that we will ultimately merge with technology as a species, and you predict this will occur quite soon, how confident are you that we will have "controls in place" so that we can allow artificial intelligence (AI) to develop without accelerating into a form which is not beneficial for us as a species?

Ray Kurzweil: First of all, to clarify my primary prediction is that we will merge with artificial intelligence. Most of the futurist artificial intelligence dystopian movies has the AI, like the Terminator, versus some brave brand of humans for control of the world, but that is not how we see it evolving now. There is not just one or two grand AI's in the world; there are billions. There are three billion smartphones—those are forms of artificial intelligence, which are accessible to about half the world's population. This number should rise to about eight billion within four or five years. Thus, AI is very much integrated and a part of who we are. These are types of brain extenders, that will go inside our bodies and brains, but this distinction is arbitrary. Even if AI is outside our body it is part of who we are. Thus, I do not see it as a separate species or as an invasion of intelligent machines 'from Mars' but we create technology to extend our own reach. We have created machines that help leverage our muscles so we can create skyscrapers, we have also created these brain extenders that enable us to access all of human knowledge with a few keystrokes.

In terms of promise versus peril, technology has always been a double-edged sword. Fire has kept us warm, cooked our food but has also burned down houses. Now we have much more powerful technologies. I wrote about this intertwining of promise and peril extensively in *The Age of Spiritual Machines,* and in Chapter 8 of *The Singularity is Near,* that looks at the deeply intertwined promise versus peril of Genetics Nanotechnology and Robotic (GNR). I am an optimist, but I do think these are valid concerns. I have written about the three stages of thought in contemplating this dilemma of technology. First, is delight as that these exponential information technologies have the potential to overcome age-old problems of poverty, environmental degradation, and disease. Then secondly, alarm at the potential that these technologies can be destructive, which we have seen technology amplify destruction throughout human history. Thirdly, where I think we need to come out, is that we have a moral imperative to continue progress, since it only this continuation of technological progress that will alleviate the continued suffering of humankind.

There is a paradigm to attend to the peril. While it is not a foolproof one, consider biotechnology for example. I mentioned some of the benefits of biotechnology, such as eradicating disease. But the same technology that could reprogram biology from cancer could also be deployed by a bio-terrorist to reprogram a benign flu virus, for example into a deadly or communicable biological superweapon. These issues were recognized forty years ago at the Asilomar Conference in California, to deal with the issue of controlling such peril. They came up with what are now known as the Asilomar Guidelines. These are ethical guidelines and strategies to keep biotechnologies safe. This conference was quite forward-thinking since, at the time, neither the promise nor the peril of biotechnology was close at hand. But now it is. We are now getting profound benefit, which is a trickle today but will be a flood over the next decade. The number of people who have been harmed by either accidental or intentional abuse of biotechnology so far is zero. This does not mean we should or can cross it off our concern list since technology keeps getting more sophisticated, so we must therefore keep reinventing these guidelines to keep it safe. Recently we just had an Asilomar Conference on AI ethics so, I am optimistic, but I think this is actually the biggest challenge for humanity of the 21st century—how do we reap the promise of these exponential technologies while also controlling the peril.

Question 3: The neural network is considered to model the human brain—do you think this is the ultimate model for the advance of our intelligence?

Answer 3: Neural networks are not actually a perfect model, since it is not actually how the brain works. Neural networks work on backpropagation, where the neural network learns by adjusting itself to the training data. It has been demonstrated that there is no backpropagation in the human brain. I have a different model, which I call the 'Pattern Recognition Theory of Mind', which does not use backpropagation but a hierarchy of modules. There is a lot of evidence now from within brain reverse engineering that suggests this is actually how the brain works. We want to achieve the kinds of flexibility and strengths harnessed in human intelligence by emulating how our human intelligence works but there many ways in which computers are far better than the human brain. A computer can reverse the alphabet for example, or compute fantastic numbers or play chess and Go far better than humans. So when a computer masters a task, it can soar past human intelligence. Some of this has to do with its very different architecture. This is also why we invented computers in the first place— to do things which humans are not very good at. Neural networks were thought to be based on how the human brain works, but now we know that they are different. Yet neural networks do have certain strengths. So I think we will pursue all these different models to create machines to extend our own thinking and be able to do things we could not otherwise do.

Question 4: What do you think about blockchain technology and how AI Technology can intersect or interact?

Answer 4: Theoretically, it is possible to have different types of currencies. The blockchain concept is where a record is kept of all transactions makes sense and has been more successful then I had expected. There is apparently a hundred billion dollars of Bitcoin in circulation, where the number of actual dollars is only about thirty times greater than this cryptocurrency, which is impressive. I have never heard of anyone actually using the blockchain to reverse engineer some transaction and resolve a dispute, nor I am sure that Bitcoin will be the currency to do that. However, there is no reason why national currencies should continue indefinitely. I think national currencies are going to weaken because it is already much a world economy in a world culture, and the generation growing up today feel as if they are citizens of the world because they are in touch with everyone. New currencies that transcend national boundaries will be inevitable.

Question 5: People seem to be becoming fearful of the loss of jobs and industries going extinct as a result of advancing AI and technology. What skills and education would you recommend to people to help make themselves 'future-proof'?

Answer 5: I have written about this. We have already eliminated all jobs several times throughout human history. How many jobs circa 1900 exist today? 38% of the population in the United States, Canada and Europe, even higher in Asia, works on farms at that time, 25% work in factories, which means about 2/3 of the population works in either factories or farms. If I were a prescient futurist in 1900, I would predict that by the year 2015, there would be only 2% on farms and 9% in factories, and this is what happened. However, it does not mean that everyone was out of work by 2015. We eliminated all these jobs at the bottom of the skill ladder to create room for new jobs at the top of the skill ladder. Now, we do not necessarily know what these jobs for the future will be exactly. This is not a great answer from a political standpoint, but this is the essence of the political problem. We have better idea about which jobs would be eliminated for our future now than we did in 1900, because our communication is better. We know that self-driving cars are a reality in the near future, but how near? We know certain jobs will be affected by this progress, but we do not know how near in the future this will occur. Consider even six or seven years ago no one knew that we would be generating millions of new ways to make money off applications on mobile phones, which is now a major part of the economy.

So what happened over the last 115 years? The American statistics (which are similar to Canada's) show that we went from 24 million jobs to 142 million jobs. Even as the population has grown, we have created many more jobs than we have eliminated. These new jobs are also of better quality, paying 11 times as much per hour in constant hours as they did in 1900, while also reducing the work year from 3000 hours to 1800 hours. Not everyone loves their job, but now there is more of a sense of getting meaning from one's work as part of their purpose and identity, which is a new phenomenon. This signals that we are moving up Maslow's hierarchy of needs, as we look for employment to provide value and meaning to our lives. Economically, within ten years these types of technologies like 3D printing, vertical agriculture, and renewable and solar energy will meet generate the material resources for a very high standard of living for all of the biological population.

Yet, we still need something to do that is meaningful, simply providing the physical things we need in life is already not enough. People find meaning in new intellectual creations, whether its music, art, literature, science or technology or new apps for mobile devices, which are different ways of expressing ourselves. So the issue is finding some work to do that is meaningful. The idea that in the future we might have tens of millions of people out of work and starving on the streets is just not going to materialize. I support Universal Basic Income (UBI), which is beginning to become a reality and will become more socially acceptable. This will allow us to have a very good life in terms of physical resources, and I think we will actually have new ways of expressing ourselves as we become smarter with brain extending technologies, which were impossible before.

Question 6: Do you think that AIs will have the necessity to create art, which helps create their identities, or will it be the end of cultural heritage and art as we know it?

Answer 6: We have always used the most advanced technology to create art. I have written about the history of music, and we have always used the most advanced technology to create music— woodworking skills in the 18th century, mechanical technology of 19th century, and today we use computers. My father was a famous musician who died in 1970, and he could only rarely hear his orchestral compositions, and that would require funding. If that opportunity came through, I would be up with him late at night using a mimeograph machine to run off scores and then finally, he could hear his composition and maybe then change a few notes or change order of passages, but it remained very difficult to make substantive changes without starting over again. Today, a student in their dorm room with synthesizer software and a MIDI [Musical Instrument Digital Interface] keyboard can create a whole Orchestra or jazz or rock band. So I think the advance of technology is only going to enhance our ability to express ourselves artistically, but it is also very important to preserve our cultural heritage because anytime we create something it is on shoulders of what came before.

Ethics of human enhancement and artificial intelligence

3.

The 'new person' contested: Atheist humanist vs. Catholic worldviews on transhumanism

Irene J. Dabrowski

St. John's University

Anthony L. Haynor

Seton Hall University

Abstract

The essay has as its main goal to foster a serious dialogue between the advocates of a transhumanist revolution, on the one hand, and Catholic critics of that movement, on the other. An argument is made that transhumanism is rooted in an atheist humanist worldview that aspires to maximum human perfectibility through the development and employment of various enhancement technologies. Drawing on the insights of the Catholic theologian, Henri de Lubac, the atheist humanist foundations of transhumanism are identified, setting the stage for a radical critique of this movement and its underlying presuppositions regarding the nature of human personhood and the human good. The challenge offered by Catholicism to transhumanism's 'new person' is unpacked, and a counter-vision of personhood, in an age when genetic engineering and artificial intelligence technologies are advancing considerably, is put forward. Not ignored in the essay is the question: 'What, if any, common ground exists between atheist and Catholic humanist perspectives on transhumanism?'

Keywords: Transhumanism, Personhood, *Homo Nouveau*, Human Enhancement, Atheist Humanism, Catholic Worldview, Henri de Lubac

Introduction

The revolution that has come to be known as transhumanism has become an object of prodigious scholarly investigation. A central concern of scholars working on this topic revolves around the implications of this nascent movement for the future of our species. Notable in this vein is the work of Don Simborg who in his recent book (2017) poses the provocative question, "What comes after *Homo sapiens*?" and then embarks on an inquiry into the possible evolution of a new human species, what he calls, '*Homo nouveau*'. While not completely discounting the role of catastrophic events or good old-fashioned natural selection as key movers in this evolution, Simborg argues that the emergence of *Homo nouveau* would in all likelihood be the result of the conscious interventions of the currently existing hominid species, *Homo sapiens*. He raises the provocative question of whether these interventions, principally in the forms of genetic engineering and artificial intelligence, can possibly produce a new form of hominid being, *Homo nouveau*, which could be created without necessarily causing the extinction of *Homo sapiens*, although that contingency, Simborg argues, cannot be completely ruled out.

While Simborg is preoccupied with the emergence of a new *biological* species (characterized by practical immortality and exponentially heightened intelligence, perpetuated by the presence of reproductive barriers vis-à-vis *Homo sapiens*), our focus here is rather on the emergence of a new *ontological* person, on the level of fundamental 'Being'. Does pronounced human enhancement represent a new type of ontological being? If yes, does this portend well or ill for human flourishing?

In attempting to come to terms with the transhumanist movement, we will fix our gaze on how it legitimizes itself as a set of institutionalized practices. What justifications are put forward to defend the ramping up of enhancement technologies? Can we discern the broad outlines of a particular 'symbolic universe' (Berger and Luckmann, 1966) that undergirds transhumanism? By this we do not refer only to the assertion that curing diseases, augmenting the quality of human life, and extending one's life are desirable goals, for these are goals to which virtually all of us, regardless of life philosophy and worldview, would subscribe. We have in mind, rather, the deeper philosophy of life and broader worldview that confers meaning on transhumanism and in terms of which adherents understand its ultimate purpose and rationale.

In this essay, we argue that the transhumanist movement is grounded fundamentally in 'atheist humanism' (de Lubac, 1983), a life philosophy that seeks human perfectibility (through the accumulation of scientific knowledge and its practical application) divorced from a supernatural source. As Kurzweil in his documentary, *Transcendent Man* (2009), put it when asked whether God exists, "well, not yet."[1] It is not the belief that human betterment can be achieved through science and technology that distinguishes atheist humanism from alternative symbolic universes. Rather, central to this worldview is the belief that human beings are capable of greater and greater perfectibility—that we can become Gods. Correlative with the god-like aspirations of our species is the sacredness attached intrinsically to life itself (Taylor, 1992). Death is seen as something to be denied or postponed, or as a challenge to live fully the one life given to us. Atheist humanism seeks human immortality in the earthly, temporal domain, by seeing the products of one's life to a significant extent as a legacy to be handed on to future generations. Transhumanism, embedded in an atheist humanist symbolic universe, can be interpreted in eschatological terms, that is, as positing an ultimate destiny for humankind, one that minimizes if not transcends pain, suffering, and limitation and brings about an earthly paradise that maximizes human flourishing. This (as we will explain in our discussion of de Lubac below) contrasts sharply with a Catholic Christian eschatology that situates paradise in the next life, in the heavenly domain, and connects it to our salvation, a right relationship with our Creator and the incarnate Christ.

Pursuant to sketching the contours of an atheist humanist defense of transhumanism, we will address the sociological question having to do with the 'carriers' of this distinctive 'package of consciousness' (Berger et al., 1974). That is to say, what forces serve to sustain it at this point in civilizational history? A very quick answer that needs to be unpacked is the rise of the digital revolution, along with the sacralization of science and technology. These forces, we will argue, serve to give the atheistic humanistic framing of transhumanism its plausibility (Berger and Luckmann, 1966) in this age. In terms closely associated with Max Weber, it is our instrumental rationality that becomes divinized (Brubaker, 1984). While the excesses of transhumanism are critiqued by some atheist humanists, this is carried out from within an atheistic humanist home base, as it were, rather than from outside of that symbolic universe.

In the next section of the essay, we will present a worldview, profoundly Catholic in its sensibility and essential logic, offering a robust challenge to atheist humanism and the latter's spirited philosophical rationale for transhumanism. Its critique is rooted in a meaning system that views human life as transitory and part of a circle of creation, a process that is part of a Providential plan. While we, humans, are made in the image and likeness of God, and thus are perfectible—intellectually, aesthetically, morally, and

---

[1] Editor's note: compare Ray Kurzweil's statement on infinity and the end of the main body of his transcribed comments in the previous chapter.

spiritually—there are, according to a Catholic worldview, limits to our perfectibility and our Creator designed us that way. That is not to say that we should refrain from drawing on our God-given potential to enhance our lives and maximize biological and mental functioning. Rather, it is that the value of life comes from our struggles with our suffering, our infirmities, our weaknesses, and our limitations. Life's meaning, from a Catholic perspective, derives fundamentally from an informed conscience and an enlightened will that energizes us to adopt a 'higher viewpoint' (Lonergan, 1973), one that enables us to transcend our weaknesses and come to see our suffering as a path to personal and spiritual growth. All of this is seen as a preparation for our eternal reunion with our Creator God, where we will experience perfectibility in all of its fullness and glory, where there will be no more ignorance, ugliness, sin, sorrow, or suffering. But, that state is neither possible in this life, nor really something that should be aspired to, in light of the Providential plan and the salvation history in which we are participating. As with atheist humanism, we will seek to identify the carriers of a Catholic vision, which has been developing organically for 2000 years. We will place considerable emphasis on the vigorous critique of atheist humanism offered by Henri de Lubac (1896-1991), a French Jesuit priest and a Cardinal of the Catholic Church, an important participant in the Second Vatican Council, and a preeminent twentieth-century Catholic theologian. While de Lubac's analysis predates the transhumanist revolution in its most recent incarnation, the lines of argument he put forward can be extended to that revolution.

We will conclude the essay with a few thoughts on the juxtaposition of the two worldviews. Areas of serious disagreement as well as points of common ground will be identified. Here we will explore the questions: Can these worldviews as they bear on the analysis and assessment of transhumanism be reconciled in any way? How so?

The transhumanist revolution: An overview

Mission

Transhumanism can be seen as a movement dedicated to "the ethical use of technology to expand human capacities" (Humanity+, 2018a). It is the stated goal of Humanity+ (an international transhumanist organization) that people "be better than well" (that is, embark on a path toward perfectibility and immortality). The transhumanist movement is thus committed to human enhancement on a grand, exponential scale. This entails increasing performance "outside the realms of what is considered 'normal' for humans" (Humanity+, 2018b).

Genetic engineering

The human enhancement that is sought as inherently desirable is generally divided into two categories. The first consists of those that are 'within' human physiology as generally understood. This category obviously includes biotechnologies capable of curing disease and repairing injury. However, the human enhancement pursued goes well beyond this limited, while highly desirable, objective. It encompasses biotechnologies—both emerging and speculative—capable of shaping the genetic structure of present generations so as to prevent the onset of diseases to which particular organisms are susceptible, that is, for which they are at discernable risk, and of future generations of organisms as well, both *in utero* and beyond (Buchanan, 2011; Doudna and Steinberg, 2017; Fukuyama, 2003; Fuller, 2011). With respect to curative and preventive medicine, nanotechnology and nanomedicine can take the form of medical implants—adaptive and intelligent—for controlling obesity, anorexia, blood sugar, seizure, and mood, and also for sending information warnings. There is also genetic screening—based on the mapping of the human genome—now capable of providing useful genetic profiles that include at-risk conditions and other traits that can be modified in the genetic line of individuals (Humanity+, 2018b; Goertzel and Goertzel, 2015; Simborg, 2017, Chapter 12). The distinction between modifying somatic and germline cells is crucial. As Simborg (2017) writes, "it is one thing to allow a person to see ultraviolet light if he wants to. It is quite another to allow him to pass on that trait to some or all of his offspring over generations" (168). Simborg (2017, 171) argues that the routinization of germline gene therapy could set the stage for the emergence of a new species, *Homo nouveau.*

It is the ultimate goal of transhumanism to achieve practical immortality (at the very least to extend the life span considerably) through enhanced genetic engineering. This has been termed the 'Methuselarity' (Goertzel and Goertzel, 2015).

AI, ASI, and electronic evolution

There are also the human enhancements resulting from the continuing development of machine technology—known generally as 'Artificial Intelligence' or AI and 'Artificial Superintelligence' or ASI (Brockman, 2015; Dormehl, 2017; Husain, 2017; Kaplan, 2017; Tegmark, 2017; Walsh, 2018; Yonck, 2017; Zarkadakis, 2017). Robots have already been developed that are capable of taking on specific tasks that have been carried out by humans, and the continuation of this automation trend is beyond question (Ford, 2016). This will lead to a decreasing need for human labor in many industries. Simborg (2017) in the chapter of his book entitled, "Electronic Evolution," draws a critical distinction between smart machines that are modeled after

Homo Sapien brains, but are capable of operating at a much more accelerated pace, and those that go well beyond the skill set of our species. Regarding the former, "we can create robots to assemble parts of automobiles, but we can't build one to safely bathe your baby or correctly sort your laundry" (Simborg, 2017, 177). Simborg argues that human doctors, for example, are far superior to computers in narrowing down options (due to their 'common sense'), while computers are superior to human doctors in functioning once the diagnostic universe has been considerably narrowed. So, he concludes that we still have a long way to go in building machines that mirror full human capability.

Disagreement over the timeline for these technologies notwithstanding, Simborg (2017, 181) expresses confidence that AI will ultimately reach a point where machines will not only simulate but exceed human intelligence to a significant degree (not to mention exhibiting greater quality control and dependability). It is not only the case that machines will simulate the human brain. According to Simborg (2017): "By using nanobots, tiny computerized robots that can be injected into the bloodstream and populate the brain, we will greatly enhance our brain understanding and capability" (182). He goes on to say that "these nanobots will allow us to wirelessly 'download' all information in our brains into a computer so that basically a computer will be able to think like a specific individual human" (182). The upshot of the singularity is that "the biological and silicon brains will be interchangeable" (182). He concludes that "a future computer will think and behave like a human, only faster, tirelessly, and with access to far greater stored information on which to base decisions" (182). Simborg argues that "we already have artificial limbs, joints, cochleas for hearing and early retina replacements for seeing. But an enhanced bionic brain gets right to the core of what makes us human" (182). According to Nick Bostrom (2016), a point will be reached when the superintelligent computer will be the main driver of the takeoff in intelligence, not humans. For Barrat (2015), what is likely to happen is that Artificial General Intelligence (AGI), in which the AI program simulates human intelligence, will become ASI, at which point the AI computer will be capable of self-awareness (personhood?) and will thus be capable of modifying its own software—it will become in effect a learning machine. Based on the work of Shanahan (2015), as interpreted by Simborg, an "AGI would have access to all human knowledge on the Internet…It would still have only average intelligence, not superintelligence. Once ASI is achieved, not only would it have access to all knowledge, but it would be far superior in intelligence in creating new knowledge, strategizing, and doing highly intelligent things we can't imagine (Simborg, 2017, 196). Kurzweil (2000; 2006; 2013) emphasizes the degree to which superintelligence involves the networking of enhanced humans that takes the form of a global brain. Then there is the distinction drawn by Markoff (2015) between AI, which mimics human capabilities (including our emotional or sentient side), and IA (Intelligence Augmentations) which extends human capabilities in vastly new directions. One can take the position, as Simborg (2017, 205) does, that "neither the most advanced robot nor a superintelligent computer will ever be considered *Homo nouveau*" (according to the strict biological definition of speciation) and still raise the question (as he and others do) as to the existential implications of these enhancement technologies. Are we becoming radically different persons?

The atheist humanist foundations of transhumanism

Human mastery and control over our evolutionary course

In his pathbreaking book, *The Drama of Atheist Humanism* (1983), originally published in 1944, Henri de Lubac offered a penetrating analysis and critique of what he termed 'atheist humanism', which he described in terms of the utopian pursuit of human perfectibility, both on the individual and collective levels. For de Lubac, this project picks up considerable steam with the 'death of God'. God is abandoned (or 'murdered'), being seen as a burden or yoke rather than a life source. This is the legacy of Feuerbach, Comte, and Nietzsche—seeing God, respectively, as a human projection, as being replaced by a religion of humanity, as within us as potential Supermen (de Lubac, 1983, Parts One and Two).

In an effort to build ourselves up, the ideal qualities (reason, knowledge, liberty, justice, love, power) once projected onto a Divine Being are now appropriated for ourselves. This zero-sum conception (when we empower God, we derogate humanity) is reframed in a Christian humanism where "Christianity does not deny man in order to affirm God. Nor does it seek a compromise between them. In fact, its revelation of God was a promotion of man..." (400). For Christian humanism must and can only profess "a lofty idea of man" (400), given the special status of humanity—being created in the image and likeness of God. Christian humanism "prepared, aroused...by a series of technical and social transformations of which previous centuries offer scarcely any examples, by progress and also by catastrophes previously unknown to us, a new man is sought today" (401), one that will "save [humans] from defeat, inviting [them] by means of the best paths to a higher success" (401-402). De Lubac argues (hardly controversially) that in most of human history, "civilization left man close to nature" (403). But in the present civilizational age, "man has...raised his head against this destiny that was weighing him down. He wants to escape the fatalities that, from time immemorial, he had learned to believe were invincible" (403).

De Lubac goes on to identify "three facts of consciousness" (403) that represent a systematic expansion of human effort and activity. The first of these is "faith in science," which de Lubac argues was "built up by experimental research and [achieved] its precision thanks to mathematics" (403). The "faith in science" has come to involve a "worship of research as being the highest ideal that can be set" (403-404). Positive science, which once "conquered only a limited field" (astronomy, the physical sciences) extended into the emerging life sciences (biology) and eventually into the human sciences (political economy, sociology, psychology). According to de Lubac, "man himself became, like external nature, the object of positive science, submitted to the same methods of investigation, and, consequently, it should be readily added, for the first time able finally to know himself, since positive science is the only rigorous type of knowledge, expelling all dreams, all imaginary visions of empty religions and philosophies" (404). For de Lubac, "this science is not only theoretical science…it is an operating science…oriented toward the possession of the world" (405). As de Lubac put it, "knowledge of the laws of the universe is, in the hands of man, a tool to be used in acting on it… Man felt destined to organize the planet with a view to its maximal output. He undertook to transform the world through his work" (405-406). So the faith in science pivoted seamlessly to a second fact—the idolatry of technique or technology, which represents the application of theoretical knowledge to practical spheres of life. De Lubac argues that "to the transformation of nature…must be added the transformation of society" (406). Here we see the development of 'social engineering'.

The third fact involves the transformation of man himself. As de Lubac put it, "one after the other, an applied biology, an applied psychology, an applied sociology were founded… through science, man was going to make himself master and possessor of the human forces. A whole 'technology' of man developed" (406-407). De Lubac cites two contemporary examples in this connection. The first is eugenics: "Thanks to progress in biology, man could now direct his own biological evolution…It would not be a question only of negative measures, like preventing certain undesirable procreation through processes such as sterilization; more must be dared: it was necessary to produce in a positive way the appearance of a superior race; and in order to do so, to specialize breeders, to refine methods of fertilization" (407). The other example was political propaganda, which is "based on a systematic exploitation of the fundamental principles of experimental psychology" (407). De Lubac argues that "the understanding of the behavioral mechanisms brings with it the possibility of manipulating them at will" (407).

De Lubac argues that the scientific mentality turned from the control of nature and society to the control of ourselves. He states: "humanity discovered itself, grasped itself as an object, [and] with a view to maximal output, it took charge of itself. It was going to forge its own destiny" (407-408). He cites Julian Huxley, who argued that humanity "substituted the possibility of conscious control of evolution for the previous mechanism of the blind chance of variation...Having attained, it thought, mastery of biological, psychic, and social phenomena, it had no doubt of being able henceforth to act on itself with the same success that it had already known in the applications of the material sciences" (de Lubac, 408). Humanity thus came to believe "in the power of rational organization... [that would] permit [it] to break with its routines and its servitudes in order to construct itself" (Huxley quoted in de Lubac, 409).

While de Lubac's analysis of atheist humanism predated what has come to be known as the transhumanist movement, we would argue that the latter represents an extension and application of the fundamental logic and worldview of the former. Clearly, transhumanism is thoroughly secular in its orientation—the Divine gives way to human control and mastery of the universe. Transhumanism is very much rooted in a faith in science and the application of scientific knowledge to nature, society, and ultimately, ourselves. It is difficult coming away from an examination of the transhumanist creed (as articulated in organizational documents) without being struck by its 'engineering' attitude. Even though the 'movement' is amorphous and variegated, it is this orientation that seems to confer upon transhumanism any coherence that it possesses.

The ethical posture of transhumanism

But, toward what ends is human reengineering and the control of species evolution (far beyond that which emerges from natural selection) directed? Does transhumanism have a well-articulated ethical philosophy? Implied at the very least in de Lubac's analysis is a preoccupation with human flourishing. Scientific knowledge is applied, and new technologies are developed in the pursuit of greater human capacity, even happiness. Nature, society, and human organisms are to be designed and redesigned based on rational principles that lead to 'progress'. We humans, by dint of our commanding intelligence are ultimately responsible for and become the driving force behind our evolution. We can and must engineer our future in the service of our continuing flourishing. As Charles Taylor (1992) framed it, according to a secular humanist worldview, life itself has the ultimate ethical value. For transhumanism, the maximizing of human capacities is an incontrovertible human good in that in doing so we live more fulfilling and productive lives, both individually and collectively.

It would be a mistake to argue that transhumanism is amoral in its fundamental posture. According to the Humanity+ mission statement, transhumanism is open to exploring the "downside of technologies" and "possible problematic and dangerous outcomes" produced by them (Humanity+, 2018b). There is a commitment in the Humanity+ community at least to human autonomy when it comes to the adoption of new biotechnologies and AI. In addition, there is a commitment to 'procreative liberty', meaning that adults should determine which traits they will possess and pass on to their progeny. Many transhumanists, while recognizing that new technologies will initially only benefit the privileged in society, assert a faith in a market process that will drive down costs and make new technologies more within reach of the majority of the population—an egalitarian dynamic they very much support. Rejected is a charge leveled against transhumanism to the effect that it supports weaker organisms being weeded out and replaced by stronger ones. The professed goal, rather, of this movement is to provide everyone with an opportunity to avail themselves of technologies that they determine will foster their flourishing. Democratic humanitarianism is affirmed, an ethical stance that opposes any totalitarian imposition of human enhancement technologies on a population or members of a population without their consent. In determining whether or not a particular technology should be encouraged many transhumanists would seem to be supportive of what has been labeled 'the proactionary principle' (Fuller and Lipinska, 2014), a prudent assessment and weighing of potential benefits and costs, factoring into the utilitarian calculus the probability estimates associated with each as well as the magnitude of possible positive and negative outcomes.

The discussion above is an attempt to identify the ultimate ends or values associated with transhumanism—human progress, human flourishing, human autonomy, human equality, democratic accountability. As we have argued, they are values that are rooted in an atheist humanist worldview. We are certainly not ascribing a monolithic quality to the transhumanist community (or to Humanity+ for that matter). No doubt there are differences of opinion when it comes to which values to emphasize, how to harmonize tensions between and among different values, and the moral judgments that are made regarding the desirability of particular human enhancement technologies. An interesting and important question is: Are conflicting judgments as to the desirability of particular human enhancement technologies the result of value conflicts or the result of differing analyses of facts and probable scenarios? We take the position (certainly provisional) that there is fundamental value consensus within the transhumanist community. This does not prevent disagreements over the desirability of particular human enhancement technologies, given that a utopian or dystopian attitude toward any given technology is, we would argue, more the result of differing empirical conclusions, as opposed to significantly different value commitments.

Carriers of atheist humanism and transhumanism

Atheist humanism and transhumanism, more recently, clearly have their roots in the movement within modern civilization toward science and technology, and away from theological foundations (Harari, 2017). As we will attempt to make clear in the next section, theology has attempted to reconcile itself to the scientific/technological age, to the engineering mentality of the current epoch. The 'rationalization' of life has been identified by Weber and many others as a major civilizational trend (see Brubaker, 1984). Contributing also to the 'plausibility' (Berger and Luckmann, 1966) of transhumanism is the 'reality' in the minds of many who live in the current period of civilizational history of a 'network' that includes not only 'people' and 'organisms' but also (perhaps even especially) 'things'. Our lives are increasingly a 'hybrid' of the 'natural,' the 'social' and the 'technological' (Latour, 1993; Pyyhtinen, 2016; Fuller, 2006; Elliott, 2015; Mitchell, 2015). It is left to individuals to determine the 'valences' to be assigned to various objects (physical, organic, artificial, supernatural) in their worlds. This 'choice' becomes a major source of their autonomy, particularly in the digital age. The options include an 'alienative' response (assigning a strongly negative valence), an 'enhancement' response (defining the object instrumentally, as a means to an end), a 'hybrid' response (seeing the object as occupying a compartmentalized space within one's life), an 'incorporative' response (defining the object as a critical part of a more encompassing self-identity), and, finally, an 'affiliative' response (constitutive of your self-identity). Our lives revolve around this complex and variegated object network, including, of course, artificial/human-engineered/technologized objects (Dabrowski and Haynor, 2017). Transhumanists point to the smartphone as being for many appendages or as extensions of themselves ('affiliative' or at least 'incorporative'). An openness to enhancement technologies needs to be put within the context of an ever-increasing comfort level with the artificial or human-built world, very much supported by the institutional structure (economic, political, educational) of society. Living in such a networked world has become for us whether we like it or not (i.e., those who adopt an alienative stance) our 'reality structure' (Berger and Luckmann, 1966) in the digital age.

A Catholic critique of transhumanism

De Lubac's critique of scientism and technological utopianism

De Lubac presents various positions on the developments noted by him. The first is to argue that in the contemporary scientific/technological age "man is biting into the forbidden fruit," that is, "usurping a role that cannot be his, of upsetting the order of the cosmos, of encroaching on the rights of God" (411).

He notes that this view is characterized by many (not only atheist humanists) as reflective of a nostalgic attachment to a form of life that has vanished forever. De Lubac presents the view of "traditionalists" that "the will to transform the world, society, and even man himself to be an even more monstrous collective revival of the crime of Prometheus. They persistently remind man then that he is a creature, that he is a part of a universe with laws that are independent of his mind just as of his will, that his first duty is to submit in all things to the real, to the object, and to respect the order established by Providence" (411). De Lubac rejects "the traditionalist refusal," arguing that "submission to Providence is not abandonment to Fatality" (414). He states: "Man was not placed in the universe as one thing among other things. Nor has he been installed there in order to enjoy it passively, as if everything had been achieved first without him. He was created, Genesis tells us, 'in the image of God'. He should then imitate him in his manner of dominating nature… through the indefinitely fruitful association of brain and hand" (414). The same, for de Lubac, applies to our dominion over society and ourselves. De Lubac advocates a "harmony of tension" (417) between the scientific/technical and the Christian minds. He states that "one will no longer conclude that one must be sacrificed to the other, that one must be condemned in the name of the other" (417). De Lubac puts forward the position that "our faith approves and encourages this effort of liberation and humanization without dictating its paths. Christianity was from the outset a 'revolt against destiny'" (420).

That having been said, de Lubac addresses the question of whether our modern Prometheus is Christian by arguing that "unfortunately, on the whole, we know only too well that the answer is No" (421). Instead of deepening and purifying Christian values (e.g. the dignity of the human person), are we not depreciating them? As de Lubac puts it, "humanity no longer sees anything today but its effort toward earthly construction. The successes or the promises of technology have gone to its head. A kind of Dionysian intoxication has seized it… In his desire for a liberation whose instrument must be technology, man goes to the point of renouncing everything that makes his condition dependent" (421-422). Everything becomes reduced to "the earthly plane," managed through an "organizing scientism" (422).

To be avoided, for de Lubac, are the illusions of the absolute as well as the illusions of the relative, of the mystical as well as of the empirical. We are charged to seek harmony between these realms, to appreciate fully how each implicates the other. In the various formulations of the religion of humanity (whether in a Comtean or Marxian guise) "real persons are immolated" with little attention given for anything but the "greatest productivity of the human animal in an all-powerful technocracy." De Lubac concludes that "it is then that Christianity, with its conception of man, comes once again to give hope to the world. Without intoxicating him with dreams, without presenting to him suspect novelties, today just as twenty centuries ago strengthened by its unchanged doctrine, by its ever-new sap, it comes to save all, to accomplish all! It is a force in the present as well as a source in the past. A force too often asleep but intact. This is due, first of all, to its realism. Christian realism is a realism of fullness. Without hiding man's misery from him, it shows him his nobility as well" (451-452). His call is to "let man, confident of divine assistance take responsibility once again for the work of the six days. Let him prolong it throughout the seventh day. May he pursue as long as this world lasts the activities of Prometheus: let him light a new fire in each century, the material basis for new human strides—for new problems and new anguish" (468). At the same time, we must "beg for the descent of the only fire without whose burning nothing can be purified, consumed, saved, eternalized" (468).

Toward a Catholic understanding of transhumanism

We have quoted at length from de Lubac because we are convinced that he has something critically important to say about our present condition as we address the ethical and spiritual dimensions of human enhancement and artificial intelligence. We are not making the argument that proponents of these technologies are ipso facto proponents of atheist humanism of the kind described by de Lubac. We are saying that those who are in the vanguard of the transhumanist movement are putting forward a worldview that places the highest value on human perfectibility, divorced from our dependence on God. De Lubac himself is very clear on this point: critical to the Providential plan is the project of human perfectibility. We, humans, are charged with the responsibility to remake nature, society, and ourselves in the service of human flourishing—to be in effect, "co-creators" with the Divine (God's fellow workers) in completing the work of creation and perfecting its harmony (*Catechism of the Catholic Church*, #307).

De Lubac's insights can be usefully applied to transhumanism. At what point do AI and gene editing technologies become problematic from a Catholic Christian point of view? De Lubac would have probably argued that the capacity to extend the species brain does not in and of itself give one ethical or spiritual pause. But, how about, for example, the search for 'master algorithms'? One could argue that the discovery of such algorithms and their application to various spheres of human life seriously undermine our autonomy, contributing to the 'immolation of real persons'. The utilization of master algorithms (Domingos, 2015) serves to objectify humanity, setting the stage for manipulation and control by societal elites. Will our enhanced capacity to predict and control human behavior through data analytics be used to promote human happiness or, rather, serve the interests of dominant groups? Can a 'democratic humanism', to which transhumanism subscribes, win out over the commandeering of machine learning technology by elites that stand to benefit from the manipulation and control of the masses?

Regarding gene editing and bioengineering, de Lubac would not have questioned the use of such technology to eliminate dreaded diseases (e.g. Parkinson's Disease, cancer) and enhance the quality of life for millions of people. However, does not a critical part of life involve our responses to our imperfections? Does not the inherent dignity and worth of each and every human being require acts of will and purpose to build on our strengths and address our weaknesses? From a Catholic Christian perspective, the genetic and operant roulette of life is very much part of the human condition. How do we make the most of the gifts we have been given by our Creator? How do we compensate for our weaknesses? The joy in life lies not in being as perfect as we can due to various artificial brain enhancements and extenders, but rather, involves how we take what we are given (the hand we were dealt) to perfect ourselves. The goal of transhumanism is to minimize struggle; while from a Catholic Christian perspective struggle is itself what gives human life ultimate meaning. De Lubac and Catholic Christianity raise important questions about what arguably is a major goal of the atheist humanist and transhumanist projects—the establishment of a paradise on earth.

Recent Christian theological scholarship on transhumanism parallels de Lubac in critical respects, although it must be pointed out that, surprisingly and regrettably, *The Drama of Atheist Humanism* was not cited in any of this work. Haughey (2014) comments that "try as we might we will not be able to produce the immortality we seek" and that transhumanism tries to establish our "dominion over our mortality" (2). Haughey (2014), similar to de Lubac, points to the eschatological imagery of transhumanism in the sense that it sees technology as a 'path to salvation'. In contrast to Haughey's more dystopian tone, Haught (2014) emphasizes the relation of transhumanism to 'he anticipatory universe'. He identifies a theological basis for arguing that humanity is continually in a state of becoming by virtue of its inventiveness. The universe is unfinished, and humanity continues to write new chapters in its evolutionary story. Cosmic transformation is to be seen as a critical part of humanity's preparation for the coming of God. Haught presents what he calls the 'sacramental option', which embraces a conservative, passive view toward technology, consistent with our profound humility in the face of the overwhelming power of Divine creation. At the opposite extreme is 'activism', which emphasizes human dominion over the earth and includes an exhortation to develop new technologies in the service of human progress. The intermediate position, what Haught calls 'promissory', sees human beings as participants in the ongoing creation of the Universe. De Lubac would in all likelihood have planted himself squarely in the promissory camp, having rejected activism as scientistic and disconnected from the Providential plan. The sacramental option would have been seen by de Lubac as too restrictive. Haught (2014) evaluates transhumanism in terms of two critical ethical benchmarks. The first is 'vitality', referring to the purposive striving of human beings in acting on their environment. Clearly, transhumanism meets this benchmark, given its strong activist bent. The second is 'subjectivity', which speaks directly to human interiority. Haught argues that transhumanism would have the effect of producing a world without subjects, given its penchant for cognitive engineering. Our minds become something to re-engineer—authentic subjectivity as a journey toward intellectual, moral, and religious conversion (see Lonergan, 1973) is seriously, if not irreparably, compromised. Human nature as it has been understood for centuries becomes passé. This sentiment is akin to de Lubac's reference to the immolation of human beings.

As interpreted by Delio (2014), Teilhard de Chardin (2004; 2008) would seem to fall into the promissory camp as well. Technology (including human enhancement technologies) is interpreted in terms of the unfolding of human consciousness in history (Hefner, 2003). Teilhard makes reference to the 'noosphere' which speaks to the emergence of a global network of collective mind (a global brain?) that has the potential, if animated by God's zest for life, to bring humanity (ultra humanity) together in love and harmony. Teilhard and Hefner share the view that technology is an expression of our self-transcendence, and that religion, rather than being external to technology, is critical to its dynamic movement. The issues raised by Delio (2014), namely, whether transhumanism leads to a lost sense of the created, the dependent, the contingent, and the finite, and whether it poses an existential threat to freedom, one that could very well take a totalitarian form, can be addressed in a more reassuring manner to the degree that we recognize enhancement technologies as part of a process of co-creation with our God. This attitude would no doubt temper radical activist tendencies in transhumanism. According to Gray (2014), the theological concept of suffering is critical if we are not to lose a sense of our finitude. A major danger of technological utopianism (to which many transhumanists succumb) is to assert that the elimination of suffering is both possible and desirable as we continue to write the narrative of our self-transcendence in history.

Bazac (2014) also raises fundamental ontological questions bearing on the transhumanist movement. What happens to the 'concreteness of the specific human'? What are the implications of the fusion of subject and object—whereby the subject becomes an object to be engineered and the object (the machine) takes on subjective qualities? Should we not treat differently the radical post-human option (the development of the intelligent robot) and the hybrid form where there is a unification of machine and the human brain? With the rise of superintelligence and genetic engineering, what happens to conscience? De Lubac, in his devastating critique of atheist humanism, points out that in its search for a 'new person' through human engineering the very idea of sin can be dispensed with. From the perspective of Christian humanism and Christian anthropology, this position is untenable. Bazac (2014) also raises the critically important issue of the political interests that drive transhumanism and the segment of society that stands most to benefit from it. Auletta et al. (2014) argue that transhumanism may very well constitute a civilizational 'discontinuity' in that it represents a radical re-engineering of human physiology, and also, with mind-uploading a 'disembodiment' of the human brain. It also represents a break with the fundamental human process of 'autopoiesis', whereby human beings transform external matter into internal self-production and re-production. This presupposes subjectivity, which is under increasing attack in the transhumanist age. The control of the brain in organizing the organism's relationship to the environment and the role of consciousness in coordinating brain functioning is seriously weakened to the degree that humans are programmed (objectified) through enhancement technologies. Human beings become enslaved to a machine or to implanted software. The brain is more than the computational machine that it is characterized as being in transhumanism. According to Auletta et al. (2014), transhumanism should be judged in terms of the degree to which the integrity of the three levels of integration—the organism's relationship to environment, the brain's control over that relationship, and the coordinating role of reflective consciousness over brain functioning—intrinsic to our humanity are respected in the age of human enhancement.

In focusing on a Catholic Christian understanding and critique of transhumanism, we leave ourselves open to the question, "But is there a non-Catholic Christian formulation"? This issue is beyond the scope of the present essay. Suffice it to say that a Catholic perspective is promissory in that it encourages a human activism that takes into full account our nature as deficient, sinful, and potentially redeemable creatures of God. A more conservative Christian horizon tends to emphasize the inherent incompatibility between the earthly and supernatural realms (reflective of a more sacramental viewpoint). A more liberal, activist Christian viewpoint tends more to equate human transformations of the universe with Godliness.

The carriers of a Catholic critique of transhumanism

As mentioned above, transhumanism has emerged in the institutional and ideational soil of the scientific and technological revolutions. An engineering ethos that exhibits a preoccupation with means-ends or instrumental rationality over substantive rationality (Brubaker, 1994), as well as the emergence of ever-thickening object networks of which the 'artificial' is a crucial component, are constitutive of this stage in civilizational evolution. Living in a "silicon" world is not something that most of us see as 'alienative'. At the very least we make our peace with it. It has the status of a facticity—we cannot wish it away (Keller and Rubenstein, 2017). Christianity (particularly of the Catholic variety) has certainly accommodated itself to some extent to the forces of science and technology, but as we have seen has sounded a loud cautionary note regarding their excesses.

The message put forward in all major Catholic magisterial documents of the past half-century is that scientific advances and technological innovations represent a positive form of human self-transcendence, both framed as an expression of the human intelligence bestowed on us by our Creator. But, as emphasized by de Lubac, the scientific sensibility (which is good) should not devolve into scientism, and an attitude of techne (which is good) should not devolve into technological utopianism. In *Gaudium et Spes* (the Vatican II document on the Catholic Church's relationship to the contemporary world) it is stated that: "Today, the human race is involved in a new stage of history. Profound and rapid changes are spreading by degrees around the whole world. Triggered by the intelligence and creative energies of man, these changes recoil upon him, upon his decisions and desires, both individual and collective, and upon his manner of thinking and acting with respect to things and to people. Hence we can already speak of a true cultural and social transformation, one which has repercussions on man's religious life as well" (#4). Science and technology are both part of the solution and part of the problem. When not subjected to the proper ethical assessment (rooted in a proper understanding of our relationship to God as well as a deep appreciation for our finitude and our personhood), humanity can go down a very dark path (Dabrowski and Haynor, 2015a; 2015b). This was clearly laid out in de Lubac's critique of atheist humanism, a line of argument that applies all too well to transhumanism.

The two most recent Popes have been crystal clear regarding the potentially disastrous consequences that can be brought about by unbridled scientism and technologism. Pope Benedict XVI (see Ertelt, 2007) warned us against designer babies, and Pope Francis (2015) has talked about technology needing to serve the goal of genuine human flourishing and personhood, and not be part of an objectification and instrumentalization of nature and human life.

The Church's stance is a promissory one—that human beings need to be active participants and responsible stewards in the ongoing creation of the universe. The tendency toward scientism and technologism needs to be tempered. While science and technology are embraced, scientism and technologism (a radical activist stance) are to be strongly rejected and resisted in the modern world. De Lubac said as much three-quarters of a century ago.

It should be mentioned that there are secularist thinkers who also caution us against the excesses of scientism and technologism. Haughey (2014) characterizes Habermas (see McLean, 2014), for example, as arguing that genetic engineering places in serious jeopardy "something as foundational as how we human beings have come to understand the meaning of our lives" (3). McLean (2014) reports Habermas questioning "whether we would want to live without a sense of moral obligation: without guilt and forgiveness, without obligation and trust" (157). Hawking (see Love, 2014), while acknowledging the beneficial effects of AI warn us against its potentially lethal consequences for the species. We have here the makings of an alliance of sorts (rooted in differing 'symbolic universes') against the dystopian effects of a radical transhumanism.

Conclusion: The two worldviews and the search for the new person

While an atheist humanism and a Catholic humanism represent two very different worldviews on the contemporary project of fundamental human enhancement, specific points of intersection should not go unnoticed. Both agree with the general premise that scientific knowledge and technological advancement are human goods and indispensable features of contemporary life, and have a constructive role to play in minimizing human disease and suffering and maximizing human functioning and well-being. This having been said, the two perspectives as they are brought to bear on transhumanism (especially in its most radical form) yield very different assessments stemming from divergent conceptions of the nature of human personhood and the ultimate purpose of life itself.

There are those who would argue that de Lubac distorts the writings of Feuerbach, Comte, Marx, and Nietzsche, collapsing them all into a grand atheist humanist vision. While it is beyond the scope of this essay to address this question, we think that de Lubac's general characterization of atheist humanism provides considerable food for thought as we attempt to grapple with the implications of the transhumanist age. It is not introducing a straw man to point out (as de Lubac did) that a new philosophical anthropology began to emerge in the nineteenth century (spearheaded by the writers to which de Lubac devotes his main attention) that has profoundly shaped our thinking about our place in the world to the present day. Their differences notwithstanding, a view of the human person began to take hold that gave primacy to our ability as a species to shape and control our destiny without being weighed down by Gods we invented and a fixed conception of nature that sharply limits our progress as individuals and as a species. Atheist humanism in its various guises defends the proposition that human beings not only have the resources to continually 'enhance' themselves through technological innovation, but also that they have the moral duty to themselves and to their species to do so. We have taken pains to identify the clear ethical foundations of transhumanism and of the broader worldview that undergirds it. Atheist humanism and its transhumanist expression embrace at their core human perfectibility, that is, the enhancement of human capacities (with cognition, decision-making, and health given primary emphasis) or 'competence' (White, 1959). In so doing, we extend the scope of human mastery to areas not imagined in civilizational history. Atheist humanism (as de Lubac would agree) and the transhumanist movement which it, in a sense, helped spawn, is well-intentioned, being sincerely committed as it is to human flourishing on a grand scale, made possible by the application of accelerated knowledge of external nature and our own nature as a species. There is a fearlessness in transhumanism's embrace of an ever-evolving, increasingly complex 'materialist' network (the result of the merging of genes and silicon), in which humans continue to occupy a central place, but in which their 'nature' is continually being reconstituted in the service ostensibly of maximal human functioning and well-being.

Catholic thought raises unsettling but necessary questions regarding atheist humanism and the transhumanist agenda that is supported by it. The response of Catholic thinkers like de Lubac, Pope Benedict XVI, Pope Francis, and the theologians whose valuable work was reported on earlier in this essay, is cautionary in its basic thrust. They pose the critically important question regarding human enhancement technologies: "What is a bridge too far?" That is to say, is heightened human competence an intrinsic good in all circumstances? Are there sensible limits to the development of these technologies and their utilization? They are certainly not referring to basic enhancement technologies that can cure diseases, reduce suffering, prevent birth defects, and restore cognitive functioning. They are referring to technologies capable of producing 'Humanity+'. What, they ask, does it mean to be human, and what is the ultimate purpose of a human life? Many answers to the question of life's purpose have been offered throughout history (Ruse, 2018). From a Catholic perspective, our humanity stems from our finitude and our limitations, and the will that we demonstrate in overcoming these deficiencies as actors in the ever-unfolding human and civilizational drama in which we all participate in partnership with our Creator God and the Incarnate Christ. It is this struggle to be wise, to do good, and to tap into a transcendent love for all of creation that confers meaning and purpose on our lives. Alarm bells go off when our subjectivity is undermined by efforts to reprogram us radically and artificially, and thus disembody (and de-spiritualize) us, or when we become absorbed, as it were, into a collective mind or global brain over which we have little or no control. To be called a subject, that is, an active, goal-oriented agent in the world is the highest compliment one can give to a human being (Smith, 2011; 2015). The human person becomes 'immolated' when 'competence' becomes an absolute and self-justifying good. These dystopian consequences, arguably, come with the territory that is 'Humanity+'.

The case can be made that genetic engineering (designer babies in the extreme), indefinite life extension, and accelerated machine learning move us out of the zone of finitude, limitation, and dependence and more into the zone of omnipotence. Does this make life more or less meaningful? Clearly, atheist humanism and Catholicism are on opposite sides of this question. For Catholicism believes that the meaning of life lies in our capacity to transcend our limitations and deal courageously with our suffering. Our finitude is seen as a critical part of the natural order of things. To perfect what we have been given (as gifts and as burdens by our Creator) is what life is all about. When that status is compromised or ceded to the forces of redesign, our very humanity is lost. What de Lubac calls for at the end of *The Drama of Atheist Humanism* is a radical recasting of what it means to be a person in light of the onslaught against the subjective person—the person who sins, repents, deals with suffering, asserts his or her authenticity in changing life circumstances, and is fundamentally at home in his or her finitude. If de Lubac were alive today he would argue forcefully for us to reclaim 'the person' that has been slipping away given the ascendance of the engineered self and the disembodied brain in an emerging transhumanist world.

As we have noted, Catholic Christianity is not the sole locus of transhumanist critique. There are scholars not generally regarded as religionists (Habermas, Fukuyama, Hawking, Simborg) who have expressed serious reservations about a Humanity+ agenda, or have at the very least suggested that we could in fact become post-Sapien, and raised the question of the desirability of this development. The larger issue (relating to 'carriers') involves an instrumental rationality that has in our era run amok (Brubaker, 1984) as well as the shift to what Sorokin (1941) calls a 'sensate' culture. As Sorokin states:

> Whereas the mentality committed to the truth of faith spiritualizes everything, regarding even matter as a mere appearance of supersensory reality, the mentality dominated by the truth of the senses materializes everything, even spiritual phenomena themselves,

> viewing the latter as a mere appearance or as a by-product of material phenomenon. Hence the general tendency of the sensate mentality to regard the world—even man, his culture, and consciousness itself—materialistically, mechanistically, behavioristically. Man becomes in sensate scientific definitions, a complex of electrons and protons, an animal organism, a reflex mechanism variety of stimulus-response relationships, or a psychoanalytical 'bag' filled with physiological libido. 'Consciousness' is declared to be an inaccurate and subjective term for physiological reflexes and overt actions of a certain kind. All the conceptions and theories predicated upon a spiritual, supersensory, immaterial reality are dismissed as a sort of superstition or ignorance or as a result of the tyranny of misused words (93-94). In such a culture, material values naturally become paramount, beginning with omnipotent wealth and ending with all the values that satisfy man's physiological needs and material comfort. Sensory utility and pleasure, as we shall see, become the sole criteria of what is good and bad. A further consequence of such a system of truth is the development of a *temporalistic, relativistic, and nihilistic mentality.* The sensory world is in a state of incessant flux and becoming. There is nothing unchangeable to it—not even an eternal Supreme Being. Mind dominated by the truth of the senses simply cannot perceive any permanency but apprehends all values in terms of shift and transformation. Sensate mentality views everything from the standpoint of evolution and progress. This comes to an increasing neglect of the eternal values, which come to be replaced by temporary, or short-term, considerations (96-97).

While Sorokin's formulation may be too stark, it does shed light on an essential difference between the two worldviews examined in this essay. Transhumanism is 'carried' by an atheist humanist philosophy that tends to be very much at home in a sensate culture. Catholicism's robust critique of transhumanism is rooted ultimately in its problematizing of sensate culture. As the transhumanist movement continues to gain momentum, it will be interesting to see if a workable alliance between Catholic humanists (who adopt an intermediate 'promissory' position towards it) and secular humanist critics (who are deeply troubled by how it, in its most extreme manifestations, devalues our humanity) can be forged. Such an alliance, one can argue, can play a meaningful role in curbing transhumanism's excesses, while at the same time being supportive of reasonable enhancement technologies that would contribute to civilizational and individual flourishing.

References

Auletta, G., I., Colage, and P. D'Ambrosio., 2014. Transhumanism critically assessed. In: J. Haughey and I. Delio, eds. *Humanity on the threshold: Religious perspectives on transhumanism.* Washington, DC: The Council for Research in Values and Philosophy, pp. 41-61.

Barrat, J., 2015. *Our final invention.* New York: St. Martin's Press.

Bazac, A., 2014. Human worth on the threshold of its technological transformation. In: J. Haughey and I. Delio, eds. *Humanity on the threshold: Religious perspectives on transhumanism.* Washington, DC: The Council for Research in Values and Philosophy, pp. 21-40.

Berger, P. and T. Luckmann., 1966. *The social construction of reality.* New York: Doubleday.

Berger, P., B. Berger, and H. Kellner., 1974. *The homeless mind.* New York: Vintage.

Bostrom, N., 2016. *Superintelligence.* New York: Oxford University Press.

Brockman, J., 2015. *What to think about machines that think.* New York: HarperCollins.

Brubaker, R., 1984. *The limits of rationality.* New York: HarperCollins.

Buchanan, A., 2011. *Better than human.* New York: Oxford University Press.

Catechism of the Catholic Church., 2003. New York: Double Day.

Dabrowski, I. and A. Haynor., 2015a. Gaudium et spes and the transhumanist movement. *The Church in the Modern World Conference,* University of St. Thomas, St. Paul, MN.

Dabrowski, I. and A. Haynor., 2015b. Gaudium et spes a transhumanistickou hnuti: Spolecensko nabozenske tech-puzzle prod vacet-prvni stoleti katolicke cirkve [Gaudium et spes and the transhumanist movement: A socio-religious tech-puzzle for the twenty-first century Catholic church]. *SALVE,* 25(4), pp. 113-129.

Dabrowski, I. and A. Haynor., 2017. Moving forward by looking back. *Annual Meeting of the Society for Applied Anthropology.* Santa Fe, NM.

De Lubac, H., 1983. *The drama of atheist humanism.* San Francisco: Ignatius Press.

Delio, I., 2014. Teilhard de Chardin's ultrahumanist worldview. In: J. Haughey and I. Delio, eds. *Humanity on the threshold: Religious perspectives on transhumanism.* Washington, DC: The Council for Research in Values and Philosophy, pp. 77-95.

Domingos, P., 2015. *The master algorithm.* New York: Basic.

Dormehl, L., 2017. *Thinking machines.* New York: Penguin.

Doudna, J. and S. Sternberg., 2017. *A crack in creation.* Boston, Houghton Mifflin Harcourt.

Elliott, A., 2015. *Identity troubles.* London: Routledge.

Ertelt, S., 2007. Pope Benedict XVI condemns genetic engineering, designer babies. Available at: < http://www.lifenews.com/2007/02/23/int-191/> [Accessed 27 June 2018].

Ford, M., 2016. *Rise of the robots.* New York: Basic.

Francis., 2015. Encyclical letter, *Laudato Si': On care for our common home.* Available: <http://w2.vatican.va/content/francesco/en/encyclicals/documents/papa-francesco_20150524_enciclica-laudato-si.html> [Accessed 22 July 2018].

Fukuyama, F., 2003. *Our posthuman future.* New York: Picador.

Fuller, S., 2006. *The new sociological imagination.* Thousand Oaks, CA: Sage.

Fuller, S., 2011. *Humanity 2.0.* New York: Palgrave Macmillan.

Fuller, S. and V. Lipinska., 2014. *The Proactionary Imperative.* New York: Palgrave Macmillan.

Goertzel, B. and T. Goertzel., 2015. *The End of the Beginning.* Humanity+ Press.

Gray, H., 2014. Putting the human into transhumanism. In: J. Haughey and I. Delio, eds. *Humanity on the threshold: Religious perspectives on transhumanism.* Washington, DC: The Council for Research in Values and Philosophy, pp. 97-110.

Harari, Y., 2017. *Homo Deus: A brief history of tomorrow.* New York: HarperCollins.

Haughey, J., 2014. Introduction. In: J. Haughey and I. Delio, eds. *Humanity on the threshold: Religious perspectives on transhumanism.* Washington, DC: The Council for Research in Values and Philosophy, pp. 1-4.

Haught, J., 2014. Transhumanism and the anticipatory universe. In: J. Haughey and I. Delio, eds. *Humanity on the threshold: Religious perspectives on transhumanism.* Washington, DC: The Council for Research in Values and Philosophy, pp. 7-20.

Hefner, P., 2003. *Technology and human becoming.* Minneapolis: Fortress Press.

Humanity+., 2018a. *Mission.* Available at: <https://humanityplus.org/about/mission/> [Accessed 22 July 2018].

Humanity+., 2018b. Transhumanist FAQ. Available at: <https://humanityplus.org/philosophy/transhumanist-faq/> [Accessed 22 July 2018].

Husain, A., 2017. *The sentient machine.* New York: Scribner.

Kaplan, J., 2016. *Humans need not apply.* New Haven: Yale University Press.

Keller, C. and M. Rubenstein., eds., 2017. *Entangled worlds: Religion, science, and new materialisms.* New York: Fordham University Press.

Kurzweil, R., 2000. *The age of spiritual machines.* New York: Penguin.

Kurzweil, R., 2006. *The singularity is near.* New York: Penguin.

Kurzweil, R., 2013. *How to create a mind.* New York: Penguin.

Latour, B., 1993. *We have never been modern.* Cambridge, MA: Harvard University Press.

Lonergan, B., 1973. *Method in theology.* Toronto: University of Toronto Press.

Love, D., 2014. Stephen Hawking is worried about artificial intelligence wiping out humanity. Available at: <https://www.businessinsider.com/stephen-hawking-on-artificial-intelligence-2014-5> [Accessed 22 July 2018].

Markoff, J., 2015. *Machines of loving grace.* New York: HarperCollins.

McLean, G., 2014. Notes on a seminar on transhumanism with Jurgen Habermas. In: J. Haughey, and I. Delio, eds. *Humanity on the threshold:*

Religious perspectives on transhumanism. Washington, DC, The Council for Research in Values and Philosophy, pp. 155-158.

Mitchell, J., 2015. *In real life: Searching for connection in high-tech times.* Berkeley: Parallax Press.

Paul VI., 1965. *Pastoral constitution on the church in the modern world—Gaudium et spes.* Available at: <http://www.vatican.va/archive/hist_councils/ii_vatican_council/documents/vat-ii_cons_19651207_gaudium-et-spes_en.html> [Accessed 28 July 2018].

Pyyhtinen, O., 2016. *More-than-human sociology: A new sociological imagination.* New York: Palgrave Macmillan.

Roden, D., 2015. *Posthuman life.* London: Routledge.

Ruse, M., 2018. *On purpose.* Princeton: Princeton University Press.

Shanahan, M., 2015. *The technological Singularity.* Cambridge, MA: MIT Press.

Simborg, D., 2017. *What comes after Homo Sapiens?* Mill Valley, CA: DWS Publishing.

Smith, C., 2011. *What is a person?* Chicago: University of Chicago Press.

Smith, C., 2015. *To flourish or destruct?* Chicago: University of Chicago Press.

Sorokin, P., 1941. *The crisis of our age.* New York: E.P. Dutton.

Taylor, C., 1992. *Sources of the self.* Cambridge, MA: Harvard University Press.

Teilhard de Chardin, P., 2004. *The future of man.* New York: Image Books.

Teilhard de Chardin, P., 2008. *The phenomenon of man.* New York: Harper.

Tegmark, M., 2017. *Life 3.0.* New York: Penguin Random House.

Transcendent man: The life and idea of Ray Kurzweil,, 2011. [Documentary] Directed by Barry Ptolemy. Ptolemacic Productions Therapy Studios.

Walsh, T., 2018. *Machines that talk.* New York: Penguin.

White, R., 1959. Motivation reconsidered: The concept of competence. *Psychological Review,* 66 (5), pp. 297-333.

Yonck, R., 2017. *The heart of the machine.* New York: Arcade Books.

Zarkadakis, G., 2017. *In our own image.* New York: Pegasus Books.

4.

Modeling moral values and spiritual commitments

Mark Graves

University of Notre Dame

Abstract

For autonomous AI to benefit humanity and avoid inadvertent harm, those AI systems must account for human values, suffering, and flourishing. Modeling human moral values and spiritual commitments enables advanced AI to incorporate those significant human dimensions into its autonomous activities and development. Those models also assist developers of less advanced AI in accounting for the moral consequences in a technical context. Moral psychology can contribute to model development as can AI capability to process and extract meaning from texts. Latent semantic analysis can model morally and religiously significant texts, such as Augustine's *Confessions* or *City of God*, and analyze them using moral values and spiritual strivings identified through psychology research. These preliminary models illustrate how scholars can use early AI tools to understand better human morality and spirituality and further develop advanced models to enable autonomous AI to contribute to human flourishing.

Keywords: Augustine, Community, Josiah Royce, Latent Semantic Analysis, Machine Ethics, Moral Psychology

Introduction

For increasingly intelligent AI systems to act beneficially for people and avoid extraneous harm, those systems should take into account human values, ways of suffering, and possibilities for flourishing. When those systems lack functional knowledge of human values and flourishing, any decisions intended to promote beneficence would be at best uninformed and unguided and likely destructive. Without knowing human suffering, AI decision-making cannot plan what to avoid, would be unable to make rational moral trade-offs in the face of unavoidable human suffering, and could not identify the suffering caused in order to learn new behaviors in the future.

For simpler AI systems, with embedded or implicitly biased knowledge, it remains important for developers and/or those selecting training data to consider the moral implications that would result from the system's actions in its foreseeable situations, as well as consider the need to constrain its operations to avoid circumstances where harms—human and other—could be significant. If one were to implement those values in terms of mathematical functions, as is often done in current AI systems, one must appropriately identify the relationship between the domain of AI activity and the function's mathematical domain without introducing hazardous moral biases (Osoba and Welser, 2017). When AI systems account for human suffering and flourishing through values embedded in its decision making, the predictions, decisions, and actions form trajectories analogous to potential spiritual paths and, therefore, may depend upon analogs to spiritual commitments. Machine ethics considers how AI may make explicit moral decisions (Asaro and Wallach, 2017). The present chapter assumes such a system would build upon current AI capabilities in problem-solving and machine learning (Norvig, 2010) and argues that natural language processing could constitute a more significant component than is generally recognized. Traditional AI systems effectively search a problem space utilizing explicit and implicit goals, and when the goals and actions affect human suffering and flourishing, those processes can have moral consequences. In addition, as explicit goals give way to learning general, socially defined intentions, those possible developmental trajectories begin to resemble developmental paths one would, in human context, consider spiritual.

For those building AI systems, explicit models can clarify both implicit moral dimensions of simpler AI systems and the explicit moral and spiritual dimensions of more sophisticated AI. When developing new, powerful technology, negligence in considering the technology's foreseeable moral implications may have negative social, legal, and historical consequences. In contrast, incorporating reasonable moral considerations into one's work can yield personal satisfaction, avoid moral and legal hazards, and contribute to personal well-being, social purposefulness, and human flourishing. Regardless of whether the moral model is explicit for the AI system or implicit for the system developer, a deeper understanding of modeling moral values and their relationship to spiritual commitments can facilitate the creation of AI systems which are more likely to benefit humanity and less likely to do harm.

Taking into account these ethical factors, this chapter argues first that modeling moral values and spiritual commitments requires both identifying those values and addressing historical and cultural challenges to developing broadly applicable models. Second, methods and findings from moral psychology focus the modeling on the relationship between self-interest and other-centered social concerns and, along with the psychological and theological study of spirituality, identify certain types of human commitments as a foundation for spirituality. The process of shared interpretation of those commitments within a community serves as a model for capturing the inherent tensions leading to human spirituality and the resultant spiritual traditions. Third, because AI systems are embodied differently than humans (especially with respect to nervous system functionality) and can process text much faster than humans, this chapter argues morally and religiously significant texts are an effective source of human morality and spirituality for AI systems. Within the AI subfield of natural language processing, semantic analysis techniques extract meaning from text. In particular, latent semantic analysis can use descriptors of moral values and spiritual commitments to measure the latent (incompletely manifested) presence of those described constructs in the text, with Augustine's *Confessions* and *City of God* used as an illustrative example to demonstrate how similar methods could be used on other significant texts. Finally, these models may not only identify human moral and spiritual strivings but also can facilitate decision making that promotes human flourishing and addresses human suffering, leading to the possibility of shared human-AI moral and spiritual development.

Identifying moral values

Human knowledge of values, suffering, and flourishing depends upon individual embodiment in addition to more general knowledge transmitted and developed through familial relations, education, social structures, and cultural values. Historically, shared values relevant to suffering and flourishing have developed and been transmitted through social structures—and their oral and written means—and identified as religious or otherwise valuable wisdom (Congar, 1967; Bellah, 2011; Narvaez, 2014). Within these cultural framings, relevant knowledge can be identified for incorporation in AI systems for general moral awareness and further adapted to meet its specific purposes as well as any required social, legal, and cultural constraints.

Major challenges to identifying and extracting shared moral values from historical religious and other social institutions and cultural norms include:

1. Historical transmission of values and the development of value-mediating traditions have been heavily intertwined with global and local political movements, diverse cultural and linguistic assumptions, and numerous historical accidents and geographic limitations.
2. Some aspects of understanding human suffering and flourishing may have been limited by continuously incomplete knowledge of human

embodiment, including unknown biological similarities between races and ethnicities, the evolutionary location of the human species, and the significant cultural and other epigenetic influences on human development.

3. The values are distributed over implicit social structures, historical practices and rituals, and millennia of written documents. These values are also embedded within diverse cultural and historical assumptions that encompass most of the human historical record. The historically more recent attempts to secularize such knowledge also increases its complexity and diverse assumptions.
4. The notion of shared values may presuppose some type of common good, which has proven an elusive construct. Both global cultural diversity and the historical record suggest the descriptor "common" frequently promotes the benefit of those in current power, and thus can contain considerable politically motivated bias.

Despite the significant challenges, failure to address them will most likely result in AI systems unable to understand human values, suffering, or flourishing, and, due to their almost guaranteed different embodiment and education, the morally blind and idiosyncratically developed systems will most certainly act in ways humans cannot predict.

Some progress has already been made to address these challenges as they also affect other scholarly, scientific, and public endeavors. First, modern and postmodern religious scholarship has identified many differences and assumptions in the historical transmission and development of values and has begun developing methods to regulate the implicit assumptions brought to those studies. Second, ongoing dialogue between science and religious scholarship is leading to a better understanding of both human suffering—by incorporating findings on neurobiological pain responses and social cognition—and human flourishing—by incorporating individual psychological and collectivist social needs. Third, the globalization and interconnection of social structures occurring due to web technologies and extensive text digitization increases access to the human historical record, in digitized form, and facilitates social collaboration across cultures. Fourth, although globalization and proliferation of advanced technology may exacerbate power discrepancies and increase cultural homogenization, the decreasing impediments to cross-cultural communication can lead to a suitable pluralistic, asymptotic approximation of shared values and generic human benefits. Even though a minimalist foundation for shared values, such as the United Nations Declaration of Human Rights, may not suffice for all human ethical needs, it might suffice as an initial platform for AI moral modeling. If so, such a platform can be extended through *in situ* cultural customization for a particular technological platform; and if not, it may at least lead to a better understanding of any remaining fundamental discrepancies.

The current chapter will primarily and initially focus on the third challenge of distributed knowledge with some focus toward contributing to the fourth challenge of shared values based upon a common good; and it will draw upon prior efforts addressing the first two challenges, especially constructive postmodern scholarship and its dialogue with the cognitive sciences (Griffin, 1993; Murphy, 1997; Gelpi, 2000; Graves, 2008). Secondarily, the first challenge, historical transmission, influences the choice of text and methods in semantic analysis; and the second challenge of limited knowledge of embodiment is revisited in the final section on human flourishing and suffering. In particular, creating computational models of human morality facilitates a quest for building AI systems capable of identifying and responding to human moral action and contributing to human flourishing while accommodating human suffering. Moral modeling also supports the possibility of advanced AI systems with more sophisticated intellectual and social capabilities developing their own morality in concert with humans. In addition to the future benefit of AI systems having the capacity to model human morality, the current development and investigation of those models can improve human self-understanding of moral reasoning and behavior with direct benefit (and beneficial feedback) to (1) current practical endeavors (including personal assistants, autonomous vehicles, and development of additional advanced AI technologies), (2) scholarly investigation of human ethics (including ethics of technology), (3) public discourse on ethics and equitable justice (including fair distribution of advanced technological resources), and (4) scientific study of morality (including models for moral psychology and social neuroscience).

Modeling moral values

Modeling moral values depends upon the ethical framework selected for representing moral activity. Three major philosophical frameworks for normative ethics are: deontological ethics, which evaluates moral action based upon rules relating to one's duty; consequentialism, which uses the consequences of one's action as the basis for moral judgment; and virtue ethics, which evaluates one's motives and character traits. Although no doubt important to model all three approaches, the current chapter will focus primarily on virtue ethics for proximate and distal reasons. In the near term, moral models would most likely be helpful in open environments where a closed set of rules or a complete model of causation would be infeasible. In the long run, if AI systems approach or surpass human-level intelligence as predicted (Kurzweil, 1999, 2005), then human intelligence might prove inadequate to predefine an appropriate set of rules or predict possible consequences (Graves, 2017b). In addition, there is growing evidence that virtue ethics may better represent human moral development because the gradual development of virtue through habit (*habitus*) formation aligns well with developmental processes of other cognitive, emotive, and social development (Green, 2012; Van Slyke et al., 2012; Narvaez, 2014; Graves, 2017a; Peterson et al., 2017). Thus it might simplify AI-human moral collaboration if both humans and AI systems built moral behaviors through habit formation processes.

The philosopher of science and technology Shannon Vallor (2016, 3-16) argues rule-based ethical theories are not useful in dealing with a future we cannot predict and that prudential (or practical) wisdom is required for moral action in a world with increasingly complex technology. Vallor (2016) identifies what she calls 'technomoral virtues' needed to respond to rapidly developing advanced technology with incompletely understood consequences. She draws upon Aristotelian, Confucian, and Buddhist ethical traditions to characterize shared conceptual resonances, acknowledging the substantial differences between approaches. Cultural differences and historical trajectories generally thwart such broad unifications, but in a rapidly globalizing context a well-considered platform for dialogue and consensus-building can play a constructive role. A challenge to interdisciplinary collaboration in machine ethics is that religious scholars can readily nuance any apparent similarity between world religions or major ethical systems while those engineering new technologies require simple models for initial explorations. Any theory developed through scholarly attention to pluralistic requirements will provide a better platform than an engineer's personal and idiosyncratic moral frame uninformed by cross-cultural dialogue. Public discourse on moral implications of AI technology can build upon a long history of ethical engagement with new technologies to extend a technological ethic to AI, but building an AI technology capable of moral engagement with humans may require a novel revision to theories previously based upon human moral behavior. Prior technologies frequently enhanced human abilities to sense or act, and some pharmaceuticals can enhance human ability to think and reason. AI not only enhances human ability to act justly or prudently—such as by synthesizing data about global inequities and waste—advanced AI systems can challenge underlying assumptions about social, emotional, and cognitive foundations of morality. Moral reasoning generally depends upon the human cognitive and empathetic ability to consider another person's perspective, and AI systems must either simulate that ability or implement an alternative way to model human norms (Briggs, 2010; Rotilă, 2016).

Moral psychology investigates the development and functioning of human behavior and mental processing in moral contexts and serves as a good foundation for investigating human moral behavior. Moral psychologists draw upon methods and theories from social, developmental, cognitive, and other areas of psychology and frequently engage with philosophical study of morality and ethical systems to develop empirical theories and models for human moral development, reasoning, and action (Annas, Narváez, & Snow, 2016; Doris, Cushman and Moral Psychology Research Group, 2010; Tiberius, 2014). With the theories heavily influenced by historical ethical systems, experiments attempt to identify factors influencing or underlying human moral behavior. Advances in neuroscience open a new avenue to examine neural correlates of mental processing in moral contexts and create models of moral mental processing consistent with observed neurological functioning and experimental findings (Sinnott-Armstrong, 2008b; Narvaez, 2014).

AI systems with moral functioning and development would require a platform different from the human brain and body, but the findings and theories developed to describe human neuroscientific foundations for moral behavior and cognition and the interweaving of those foundations with the person's cultural and historical context can also inform an analogous platform for AI systems. The mapping between neuroscientific foundations for human morality and a technological platform for AI morality is simpler than one might assume because both moral psychology and AI research share a dependence and engagement with cognitive sciences, including cognitive neuroscience research. The philosopher Walter Sinnott-Armstrong (2008a, 2008b) identifies contributors to the ongoing grounding of moral psychology within cognitive science and neuroscience, and AI research has historically benefited from engineering models and processes derived from neuroscience and cognitive psychology (Gardner, 1985). Creating cognitive models capable of supporting human and/or AI moral reasoning, action, and development would enable fruitful dialogue between moral psychologists and AI developers. Many scientific, technological, and sociological problems remain, but the goal to develop cognitive models for human moral functioning (or development) and moral AI systems would define a fairly well-scoped research program on par with numerous successfully completed scientific initiatives. Modeling the major ethical systems and their differences would result in a significant scientific, scholarly, and public achievement with far-reaching impact. A foundational and more tractable precursor goal is to create models that capture the human capacity to have a single system of moral values or ethical framework.

The moral psychologists Frimer and Walker (2009; Frimer et al., 2011; Walker, 2013) build upon prior work in moral, religious, and spiritual exemplars (Walker, 1999) and just, brave, and caring moral exemplars (Walker and Hennig, 2004) to model interactions between two dimensions of moral behavior: agentic (self-interested) and communal (centered on the concerns of others). They found that for the general population, moral behavior generally associates with communal rather than agentic concerns, but among moral exemplars the disparate motivations toward agentic or communal concerns generally integrate as agentic concerns reconcile in the service of communal ones. Therefore, in addition to modeling agentic and communal concerns, a model capable of understanding people with exemplary moral functioning would need to identify the additional integrative relationship between personal and communal concerns which otherwise appear to operate independently in the general population. Although there are exceptions, modeling the moral values of individuals and their effect occurs most explicitly in moral psychology, while shifting attention to the ways communal values influence the person occurs more predominately in the study of spirituality. A complete model of moral values, especially as it relates to human-AI interactions, requires modeling of individual and communal values as well as how exemplars would integrate their behavior and personal motivations within a communal context. Although exemplars comprise only a small fraction of the population, they have a large impact on defining normative moral behavior. Individuals such as Mother Theresa, Martin Luther King, Jr, other Nobel Peace Prize recipients, and similar historical moral and religious leaders exemplify moral behavior in their context and help to define the normative dimension of human moral behavior.

Capturing individual agentic values is well studied. The psychologist Robert Emmons (1999) identifies several strivings a person might pursue as ultimate which he and other psychologists have found empirically to orient a range of human purposeful activity. Strivings include achievement, power, intimacy/affiliation, spiritual transcendence, and generativity (for example, the prosocial creation of legacy). Modeling these strivings would capture aspects of human moral values that almost certainly influence decisions in a moral context. For an AI system to model such strivings, it must have the construct of striving itself, know what strivings are possible for humans (at least with respect to the context in which the AI system operates), and have the capacity to associate particular strivings with individuals or populations, possibly varying over particular types of contexts. In addition, the model would need to capture those strivings within a communal context where individual values occur in diverse synthesis with communal values. Investigation into human spirituality can serve as a foundation for modeling the communal context of moral commitments.

Modeling spiritual commitments

Communal moral values depend upon commitments that one can study within the context of spirituality. As a working definition, spirituality is the experience of striving to integrate one's life toward the ultimate value one perceives, and that ultimate value is mediated through a tradition and its associated communities. The Protestant theologian Paul Tillich (1956) characterized a person's relationship with God in terms of their Ultimate Concern, and the scholar of spirituality Sandra Schneiders (2005) argues that spirituality refers to the experience of moving toward some ultimate value (or horizon, beyond which one cannot perceive) and integrating that movement into one's lived experience. A focus on Ultimacy loosely synthesizes many theological aspects from the world's religions, and the focus on integrative experience toward Ultimacy can characterize most associated spiritual paths (to a degree sufficient for an initial model). The Roman Catholic theologian Yves Congar (1964, 1967) distinguishes a *tradition* (like Christianity) from its cultural manifestations through its *traditions* (like Protestant denominations or Roman Catholicism). The Protestant philosopher of religion Josiah Royce (1913) identifies the significance of community to continually interpreting the tradition and its collective spirituality through the lives of its members. To avoid overreliance on Christian categories, drawing upon Mahāyāna Buddhism (Williams, 2009), *sangha* could be compared to community, *dharma* to the path embedded within the Buddhist tradition, and enlightenment and *nirvana* for all beings as the *Bodhisattva* commitment.

Community generally refers to a socially cohesive group, and as the term is used here, implies the group has a common value or concern. Royce identifies significant shared features, such as memories, hopes, and lives, and defines a certain type of community, called a community of interpretation, where members interpret their common commitment one to another. As I interpret Royce's understanding of community to those sharing (at least some minimal level of) a commitment to an interest in AI spirituality, something is created, which Royce's intellectual predecessor Charles S. Peirce calls an *interpretant*, which cannot be reduced simply to what Royce, you, or I separately understand. With sustained commitment, those interpretants build upon each other, and Graves (2008, chap. 5; 2009) argues leads to an emergent type of proto-spirituality. For Royce, when that community of interpretation occurs with memory of the life of Christ, and hope of eschaton (e.g., heaven), then the spirituality is Christian spirituality and occurs with the working of the Holy Spirit. Congar would emphasize the temporal aspects with his understanding of tradition, and Theravada Buddhists would somewhat comparatively attend to the life of Buddha as foundational memory and emphasize the trajectory as a path toward enlightenment and *nirvana*.

Modeling spirituality may require identifying the predominant social structure(s) within which the shared memories, hopes, and ongoing interpretive process occur. Shared memories include information about foundational events and major decisions or divisions along the historical path that would affect the cultural interpretations or traditions created. The forward-looking shared hope, vision, or concern almost certainly cannot be easily and fully characterized—otherwise there would be no need for a diverse community to commit to continuing interpreting it—but one can model that uncertainty itself to capture the diverse interpretations of historical and contemporary agents. The philosopher of science Michael Weisberg (2013, chap. 2) argues for a distinction between mathematical and computational models because, for computational models, the process of computation itself is significant in the modeling rather than just the mathematical structure. One could imagine representing communal memories analytically (e.g., as logical propositions), but modeling future-oriented hope may require a simulation itself as representation. In such cases, further characterizing the ambiguity may be an essential aspect of the model. For Christianity, one may consider God as Ultimate Concern and one's relationship with God as Ultimate Value, which one strives to develop and strengthen as one's spiritual path. However, the ambiguity in how to relate the salvific path and the eschatological hope lingers as fundamental differences between Protestant, Roman Catholic, and Eastern Orthodox traditions (in terms of justification and sanctification); and for Buddhism, the uncertainty with how one relates the interconnectedness of all beings with one's trajectory yields differences between Theravada and Mahāyāna traditions. Capturing the tension between doctrinal interpretations is essential to understand fully the interpretations and their underlying referents. For AI systems to act responsibly in religion-influenced contexts—including end-of-life care, human enhancement, or cross-cultural diplomacy or conflicts—they must capture the inherent tension in diverse religious interpretations. Otherwise the systems cannot understand the internal conflicts humans have with such scenarios. Like most scientific models, the areas of focus and granularity depend upon the task at hand, and a flexible modeling framework or paradigm would support a variety of measures.

An initial model for human morality can capture moral values as strivings within the community. Within that context, one can focus on the interpretations shared over time within traditions or other long-lived communities that have commitments aligned with the moral values to be modeled. In historical communities or traditions, one way to extract the interpretations is from texts considered morally or spiritually significant.

Semantic analysis

One could develop a moral and spiritual model in several ways, including programming, extending a user model based upon online social activity, extracting from interactions within a social network, analyzing written text for latent values or commitments, direct user dialogue, or other observational or interactive methods. People generally build upon their embodied capacity for social interaction, enculturation, broad education, and communication with others to develop a general foundation for their own and other's morality and spirituality. A sufficient initial foundation for building computational moral and spiritual models is to analyze a variety of morally and spiritually representative texts. There are several reasons to begin with text: communication between people appears significant for spirituality; written text has proven significant for transmission of values and commitments across generations; text analysis requires less context then behavioral observation; it has ready application to online services, such as social media, and to applying moral models in casuistry or case-based legal analysis; and (written or spoken) text would be an aspect of direct communication which may prove essential for future AI-human interactions. For humans, text might not be the best place to begin because we frequently behave differently than our education and reading in ethics might teach us, due in part to neurological differences in the storage of text semantics and planning of possible actions. The embodiment of human cognition closely links human perception to sense organs and human action to motor function. One learns to identify and form letters automatically as one learns to read and write, but learning to behave morally depends upon repeatedly acting morally. However, AI systems do not have that limitation. To a certain extent, a major goal of AI research is to make computers behave other than by just following written instructions. So, for AI systems, semantic analysis may drive moral behavior more directly than analogous reading would do for humans.

Moral and spiritual communication within humanity spans several millennia, and texts since the beginning of the Axial Age (approx. 8th century BC) appear pivotal for understanding contemporary moral and spiritual development (Bellah, 2011). The range of moral values and spiritual commitments latent within one's culture frequently have an original source (or written characterization)—or if common across cultures, may have numerous characterizations. Although humans may acquire that latent knowledge implicitly through numerous cultural sources, and no individual person has the capacity for in-depth knowledge across the range of human culture, computational methods may compensate for their limited cultural embeddedness and different embodiment through their capacity for rapid processing of large quantities of text. In addition, computational text analysis is analogous to formal human education of ethical systems, which typically includes study of morally (and/or religiously) significant texts.

Within computational text analysis and the AI subfield of natural language processing, the area of semantic analysis focuses on extracting meaning from text. Although valuable to examine texts for keywords or word frequencies, semantic analysis extracts more meaningful information by creating a mathematical model of the text semantics—often by modeling the co-occurrences of words—and then examining the model for insights into the text's meaning. This approach to semantic analysis depends upon an associationist and distributional theory of meaning. Building upon Wittgenstein's argument that the meaning of a word lies in its use in language (Wittgenstein, 1958, secs. 80, 109), the linguist John Firth (1957) characterizes the meaning of a word as dependent upon the words with which it is in frequent and habitual company, and thus, the associations between words define meaning. To model those associations, the mathematical linguist Zellig Harris (1968) identified and developed the distributional hypothesis: because words with similar meaning have similar contexts, then words with similar patterns of association in a sufficiently large sample of language have similar meaning.

One classic method for semantic analysis is latent semantic analysis (LSA), which has also been shown to have some psychological plausibility (Landauer et al., 2007). LSA computes semantic similarity between texts by translating them into mathematical representations (vectors) then comparing those representations to calculate a metric of their semantic similarity (typically, the cosine of the angle between vector representations). A typical use of LSA within psychology is to project a vector-representation of text onto a fixed-dimension semantic space created from a large corpus (using singular-value decomposition and retention of primary components), often a specific 11 million word corpus of first-grade through first-year collegiate readers (called 'TASA'), which has proven a robust space in practice (Landauer et al., 2007, 69). Reimer et al. (2012) use that approach to examine moral mental schemas among humanitarian exemplars comparing moral identity interview narratives (McAdams, 1997) with moral and religious descriptors derived from Walker's (1999) investigation of moral, religious, and spiritual exemplars and Walker and Hennig's (2004) empirical study of just, brave, and caring moral identities.

Ideally, I would now demonstrate semantic analysis against a corpus of significant moral spiritual, and religious texts and perform comparisons against theory-driven collections of them as part of an empirical investigation of human morality, religion, and spirituality. Unfortunately, such a prepared corpus is not yet available, and proper analysis and discussion of multiple historically significant texts would exceed the scope of this chapter. Instead, I describe example measures that could form moral and spiritual dimensions of an initial model based upon two classic books by Augustine. The example illustrates how similar techniques could be used on additional dimensions and other significant moral, spiritual, and religious texts.

The highly influential, fifth-century Christian writer Augustine of Hippo's major works include his *Confessions*, the first Western autobiography, and *City of God*, a cornerstone of Roman Catholic and Protestant views on the relationship between Christianity and the world. Using Reimer et al. (2012) LSA approach and descriptors on modern English translations of those books, both books were found primarily religious in comparison to just, brave, and caring moral descriptors. The cosine similarity scores for the probes also indicated that just and brave were comparably emphasized while caring was emphasized less. (Confessions: religion cos=.32, just cos=.25, brave cos=.26, caring cos=.17; City of God: religion cos=.32, just cos=.23, brave=.24, caring cos=.15). (Cosine scores generally range from approximately 0 to a maximum of 1, with higher scores indicating greater similarity.) In comparing the two texts to each other, they appear equally religious but *Confessions* appears to deal with moral issues slightly more than *City of God* (cosine difference for each of just, brave, and caring descriptors between the two books is 0.02, which is small but not trivial given the amount of text compared).

To examine other dimensions of the text, the analysis uses a similar approach to Reimer et al. and creates descriptors based upon Emmon's (1999) strivings, described above, to identify strivings in text. (Appendix A of this chapter lists the striving descriptors from Emmon's (1999) example sentences in his coding manual Appendix B with a slightly revised categorization based upon his and his students' subsequent work.) Analyzing those descriptors against both books (with respect to TASA semantic space) finds the greatest striving similarity scores for spiritual transcendence and the least for power striving (Confessions: ST cos=.51, power cos=.16; City of God: ST cos=.50, power cos=.18), which was generally unsurprising given their topics. Comparing the two books, the greatest differences were with striving identity greater in *Confessions* (cos=.36) than *City of God* (cos=.32) and self-presentation lower in *Confessions* (cos=.22) than *City of God* (cos=.26). Since *Confessions* is the prototypical autobiographical confession, it is interesting that LSA captured both higher striving for identity and lower self-presentation than Augustine's other work, though a broader comparison with additional works by Augustine would be needed to confirm that interpretation.

LSA can also be used to analyze sections within the texts. *Confessions* consists of thirteen chapters, which have been well studied by scholars. Figure 4.1 shows the cosine similarity score across the thirteen chapters for six descriptors: the religious descriptor; just, brave, and caring moral descriptors; and the relatively higher-valued spiritual transcendence and identity striving descriptors, indicted by the different lines. The peaks indicate greater similarity for a chapter to the descriptor, and the valleys indicate less similarity. The highest cosine score is for spiritual transcendence in chapter 1, which includes several introductory prayers of praise, and the lowest score of the six descriptors used in the figure is caring in chapter 11, which begins Augustine's examination of time and creation relying on the Biblical book of *Genesis*. One can also examine the cosine scores to identify patterns across chapters and descriptors: most of the descriptors have relatively minor fluctuations between chapters except for the religious descriptor, which has a precipitous drop in cosine value between chapters 9 and 11. Scholars and other readers note the difference in the chapters of *Confessions* between the first nine autobiographical chapters, with chapter 9 describing the death of Augustine's friends and mother; chapter 10, which examines memory and the writing of *Confessions* itself; and chapters 11-13, which begin to explain *Genesis*. The sharp drop in cosine value between chapters 9 and 10 for the religious descriptor illustrates the shift in religious theme noticeable between those chapters, while the relative increase in cosine between those two chapters for all other descriptors indicates the continuity or moral and spiritual aspects in Augustine's explanation. Given the scholarly debate about the coherence of the book as a whole (Kotzé, 2006), we take the mean cosine across chapters 1-9 and chapter 11-13 and note that there is no difference between those two parts for the justice descriptor (cos=.24) and that the largest difference occurs for the bravery descriptor (chap 1-9 cos=.26, chap 11-13 cos=.20, with chap 10 cos=.28). This comparison of scores indicates coherence for the justice theme and differences in themes of bravery, perhaps leading to distinctions in how engaging the text would be with respect to courage.

Figure 4.1: LSA cosine similarities between moral, religious, and striving descriptors and the chapters of Augustine's Confessions

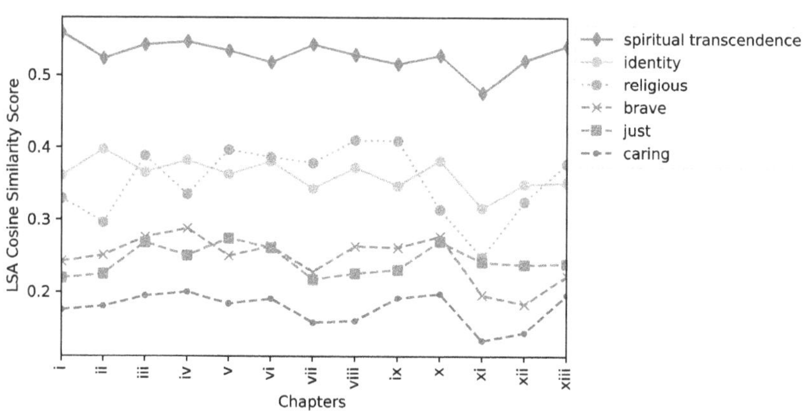

LSA can identify latent meaning relevant to morality and spirituality that may not be readily apparent to a human reader (or author). Scholars can use these techniques to explore various questions within an author's text. Preliminary findings suggest applying such techniques to the writings (or interviews) of moral and spiritual exemplars may yield fruitful insights not readily available without a high level of scholarly familiarity and expertise with the text. Computational methods may also enable comparative work that would otherwise exceed the mental resources of all but the most seasoned human scholars.

In addition to LSA, other semantic analysis techniques can characterize additional approaches and dimensions for moral and spiritual models. The Linguistic Inventory and Word Count (LIWC) system (Tausczik and Pennebaker, 2010; Pennebaker et al., 2015) uses fairly standard word counting techniques with approximately 80 psychologically validated and linguistic word categories to calculate category text frequencies, which can model explicit themes within the text. The technique of topic modeling treats each document (or section of text, such as a paragraph) as comprised of a small number of topics then uses probabilistic methods to estimate the topic as a probability distribution of words (Blei, 2012), and thus enables 'unsupervised' analysis of text semantics. The development of more computational methods for semantic analysis is an ongoing research endeavor, with significant recent progress in neural network models (Goldberg, 2016).

Fruitful scholarly investigations may continue to increase available texts, refine the dimensions modeled, extend the semantic analysis techniques, and pursue comparative analyses. Social scientists and other investigators can use these continuously improved models to better understand moral and spiritual dimensions of exemplary and other contemporary populations, and AI developers can use the refined models and improved understanding of human morality and spirituality to develop systems that examine and respond to human moral values and spiritual commitments. As these systems deploy and enable the development of new methods and experimental paradigms, that new knowledge can incrementally ratchet up additional scholarship, scientific analysis, and technology development, feeding forward in a self-improving cycle. Extending AI from systems with the capacity to analyze text for moral and spiritual constructs to those that can act intentionally based upon their moral knowledge will certainly be a challenging endeavor, though perhaps no more difficult than the growth required from human children who can comprehend moral stories, fables, and situations yet must learn to apply moral principles and virtues practically (prudentially) in the adult world. If AI systems do reach the stage they can make autonomous and agentive decisions, then having semantic analysis and other AI techniques in place for moral and spiritual discourse may focus joint AI-human activity on communal benefits and lead toward human (and perhaps mutual) flourishing and exemplary behavior.

Human flourishing and suffering

Human flourishing results from a specific kind of purposeful happiness or well-being, in Greek *eudaimonia*, which is distinct from the subjective feeling of happiness (called *hedonia*) that refers to a feel-good, happy life. A strict definition of the hedonic life is captured by the presence of pleasure and absence of pain, though a broader characterization can include a positive attitude and life satisfaction (Kahneman, Diener and Schwarz, 1999; Diener, 2009). Eudemonic well-being, in contrast, is fairly independent of subjective feelings of happiness, which are heavily influenced by circumstances, and relates instead to striving toward a committed purpose in life (Deci and Ryan, 2008). For Aristotle, eudemonia refers to both the end goal (*telos*) as well as the resulting kind of happiness. For the Christian philosopher Thomas Aquinas, the ultimate goal is to know and love God, from which eudaimonic happiness would result.

Pain generally refers to the unpleasant experience neurobiologically associated with tissue damage, though it can readily be caused by other illnesses or injury as well as isolating social rejection or loss (Fields, 2007; Eisenberger, 2012). Suffering typically refers to subjective responses to pain and can be exacerbated by anticipation or rumination. Accounting for suffering is very important to consequentialist ethical frameworks, and all major religions acknowledge the existence and inevitability of pain and the resultant suffering with Buddhism explicitly addressing the suffering by identifying its cause, emphasizing its transitory and conditioned existence, and prescribing a path toward its cessation. Christianity has had a more ambiguous relationship with suffering since painful consequences could result from sinful actions or from morally correct actions, and pain can have potential cathartic effects, but many modern Christian theologians would emphasize the greater relevance of how one responds to suffering given its inevitability and acknowledge the importance of eliminating excess suffering (Ryan, 2011).

One complexity identified within both Buddhism and Christianity as well as in psychology is that avoiding pain and suffering generally leads to more suffering (Chawla and Ostafin, 2007). For example, avoiding a minor medical procedure may lead to greater illness. Buddhism explicitly addresses suffering through its 'Middle Way' practices, and Christianity's appropriation of Aristotelian virtue ethics emphasizes the necessity for virtue to mediate between contrasting possible vices. Both a foolhardy approach towards danger and a cowardly evasion of danger could create more suffering than a courageous response might do. Creating a model capable of reducing excess suffering (and minimizing unnecessary increases in suffering) requires a model that supports mediating between conflicting actions and motivations. Although a rescue robot that always moves toward the injured is simpler to design, that would prove a short-sighted strategy in an environment with dangers to the robot, many injured people, and limited resources. With respect to semantic modeling, Walker's bravery measure might constitute only a foundation for courage and also require modeling measures to incorporate explicit foolhardy or cowardice components.

Modeling dimensions of moral and religious texts is a first step toward a more thorough representation of the spiritual commitments that orient moral behavior. More complex models may represent whether one develops the character traits to pursue a path toward a concern, describes the necessary behaviors as a collection of rules, or evaluates consequences of actions with respect to some concern. The trajectories implicitly or explicitly define a perspective on human flourishing, and its relationship with pain and suffering, and AI developers can benefit from drawing upon centuries of investigation into how those themes relate. Although initial models will most likely have limited perspective on human values, commitments, and paths to flourishing, the incremental improvement of those models contributes to a better understanding of the human condition and may contribute significantly to flourishing for both humans and AI systems.

Conclusion

Human values broadly held depend upon and support shared commitments typically identified and maintained by spiritual and other meaningful beliefs and practices generally associated with religious and other significant cultural institutions. Capturing the values and commitments embedded within those human social systems enables explicit models that AI systems can utilize to benefit humanity and increase human and AI flourishing. A fruitful initial approach to modeling human morality and spirituality can use semantic analysis tools to extract meaning from moral and religious texts. These models can represent moral values and spiritual commitments and support activity oriented toward human flourishing.

References

Annas, J., D. Narváez, and N.E. Snow, eds., 2016. *Developing the virtues: Integrating perspectives.* New York: Oxford University Press.

Asaro, P. and W. Wallach., 2017. Introduction: The Emergence of Robot Ethics and Machine Ethics. In: W. Wallach and P. Asaro, eds. *Machine ethics and robot ethics.* New York: Routledge, pp. 1-18.

Bellah, R. N., 2011. *Religion in human evolution: From the Paleolithic to the Axial Age.* Cambridge, MA: Belknap Press of Harvard University Press.

Blei, D. M., 2012. Probabilistic topic models. *Communications of the ACM,* 55(4), pp. 77-84.

Briggs, G., 2010. Machine ethics, the frame problem, and theory of mind. In: *Proceedings of the AISB/IACAP World Congress.*

Chawla, N. and B. Ostafin., 2007. Experiential avoidance as a functional dimensional approach to psychopathology: An empirical review. *Journal of Clinical Psychology,* 63(9), pp. 871-890.

Congar, Y., 1964. *The meaning of tradition.* New York: Hawthorn Books.

Congar, Y., 1967. *Tradition and traditions: An historical and a theological essay.* New York: Macmillan.

Deci, E. L. and R.M. Ryan., 2008. Hedonia, eudaimonia, and well-being: An introduction. *Journal of Happiness Studies,* 9(1), pp. 111.

Diener, E. ed., 2009. *The science of well-being: The collected works of Ed Diener* (Vol. 37). Springer Science and Buisness Media.

Doris, J. M. and Moral Psychology Research Group., ed. 2010. *The moral psychology handbook.* New York: Oxford University Press.

Eisenberger, N. I., 2012. The neural bases of social pain: Evidence for shared representations with physical pain. *Psychosomatic Medicine,* 74(2), pp. 126-35.

Emmons, R. A., 1999. *The psychology of ultimate concerns: Motivation and spirituality in personality.* New York: Guilford Press.

Fields, H., 2007. Setting the stage for pain: Allegorical tales from neuroscience. In: S. Coakley and K. K. Shelemay, eds. *Pain and its transformations: The interface of biology and culture.* Cambridge, MA: Harvard University Press.

Firth, J., 1957. A synopsis of linguistic theory 1930-1955. In *Special Volume of the Philological Society.* Oxford: Oxford University Press.

Frimer, J. A., L.J. Walker, W.L. Dunlop, B.H. Lee, and A. Riches., 2011. The integration of agency and communion in moral personality: Evidence of enlightened self- interest. *Journal of Personality and Social Psychology,* 101(1), pp.149-163.

Frimer, J. A. and L.J. Walker., 2009. Reconciling the self and morality: An empirical model of moral centrality development. *Developmental Psychology,* 45(6), pp. 1669-1681.

Gardner, H., 1985. *The mind's new science: A history of the cognitive revolution.* New York: Basic Books.

Gelpi, D. L., 2000. *Varieties of transcendental experience: A study in constructive postmodernism.* Collegeville, Minn.: Liturgical Press.

Goldberg, Y., 2016. A primer on neural network models for natural language processing. *Journal of Artificial Intelligence Research,* 57, pp. 345-420.

Graves, M., 2009. The emergence of transcendental norms in human systems. *Zygon,* 44(3), pp. 501-532.

Graves, M., 2016. *Mind, brain and the elusive soul: Human systems of cognitive science and religion.* Routledge.

Graves, M., 2017a. Habits, tendencies, and habitus: The embodied soul's dispositions of mind, body, and person. In: G.R Peterson, ed. *Habits in mind: Integrating theology, philosophy, and the cognitive science of virtue, emotion, and character formation.* Leiden: Brill.

Graves, M., 2017b. Shared moral and spiritual development among human persons and artificially intelligent agents. *Theology and Science,* 15(3), pp. 333-351.

Green, B. P., 2012. Teleology and theology: The cognitive science of teleology and the aristotelian virtues of techneand wisdom. *Theology and Science,* 10(3), pp. 291-311.

Griffin, D. R., 1993. *Founders of constructive postmodern philosophy: Peirce, James, Bergson, Whitehead, and Hartshorne.* Albany, NY: State University of New York Press.

Harris, Z., 1968. *Mathematical structures of language.* New York: Interscience.

Kahneman, D., E. Diener, and N. Schwarz., 1999. *Well-being: The foundations of hedonic psychology.* New York: Russell Sage Foundation.

Kotzé, A., 2006. The puzzle of the last four books of Augustine's confessions: An illegitimate issue? *Vigiliae Christianae,* 60(1), pp. 65-79.

Kurzweil, R., 1999. *The age of spiritual machines: When computers exceed human intelligence.* New York: Viking.

Kurzweil, R., 2005. *The singularity is near: When humans transcend biology.* New York: Viking.

Landauer, T.K., D.S. McNamara, S. Dennis, and W. Kintsch, eds., 2007. *Handbook of latent semantic analysis.* Mahwah, N.J.: Lawrence Erlbaum Associates.

McAdams, D. P., 1997. *The stories we live by: Personal myths and the making of the self.* New York: Guilford Press.

Murphy, N., 1997. *Anglo-American postmodernity: Philosophical perspectives on science, religion, and ethics.* New York: Routledge.

Narvaez, D., 2014. *Neurobiology and the development of human morality: Evolution, culture, and wisdom.* New York: Norton.

Russell, S. J. and P. Norvig., 2010. *Artificial intelligence: A modern approach.* 3rd edition. Upper Saddle River, NJ: Prentice Hall.

Osoba, O. and W.I.V. Welser., 2017. *An intelligence in our image: The risks of bias and errors in artificial intelligence.* Santa Monica, CA: Rand Corporation.

Pennebaker, J.W., R.L. Boyd, K. Jordan, and K. Blackburn., 2015. The development and psychometric properties of LIWC2015. *UT Faculty/Researcher Works* [online]. Available at: <http://hdl.handle.net/2152/31333> [Accessed 8 July 2018].

Peterson , G.R., J.A. Van Slyke, K.S. Reimer, and M.L. Spezio., 2017. *Habits in mind: Integrating theology, philosophy, and the cognitive science of virtue, emotion, and character formation.* Leiden: Brill.

Reimer, K.S., C. Young, B. Birath, M.L. Spezio, G. Peterson, J. Van Slyke, W.S. Brown., 2012. Maturity is explicit: Self-importance of traits in humanitarian moral identity. *The Journal of Positive Psychology*, 7(1), pp. 36-44.

Rotilă, V., 2016. The impact of theory of mind over ethics and law: few arguments. In: *European Proceedings of Social and Behavioural Sciences, WLC 2016: World LUMEN Congress.*

Royce, J., 1913. *The problem of Christianity: Lectures delivered at the Lowell Institute in Boston, and at Manchester College, Oxford.* New York: Macmillan.

Ryan, R., 2011. *God and the mystery of human suffering : A theological conversation across the ages.* New York: Paulist Press.

Schneiders, S. M., 2005. Approaches to the study of Christian spirituality. In: A. Holder, ed. *The Blackwell sompanion to Christian spirituality.* Oxford: John Wiley and Sons, pp. 15-33.

Sinnott-Armstrong, W. ed., 2008a. *Moral psychology, Volume 2: The cognitive science of morality: intuition and diversity.* Cambridge, MA: MIT Press.

Sinnott-Armstrong, W. ed., 2008b. *Moral psychology, Volume 3: The neuroscience of morality : emotion, disease, and development.* Cambridge, Mass.: MIT Press.

Van Slyke, J. , G. Peterson, W.S. Brown, K.S. Reimer, and M.L. Spezio., 2012. *Theology and the science of moral action: Virtue ethics, exemplarity, and cognitive neuroscience.* New York: Routledge.

Tausczik, Y. R. and J.W. Pennebaker., 2010. The psychological meaning of words: LIWC and computerized text analysis methods. *Journal of Language and Social Psychology*, 29(1), pp. 24-54.

Tiberius, V., 2014. *Moral psychology: A contemporary introduction.* New York: Routledge.

Tillich, P., 1956. *Dynamics of faith.* New York: Harper.

Vallor, S., 2016. *Technology and the virtues: A philosophical guide to a future worth wanting.* New York: Oxford University Press.

Walker, L. J., 1999. The perceived personality of moral exemplars. *Journal of Moral Education*, 28(2), pp. 145-162.

Walker, L. J., 2013. Exemplars' moral behavior is self-regarding. *New Directions for Child and Adolescent Development*, 142, pp. 27-40.

Walker, L. J. and K.H. Hennig., 2004. Differing conceptions of moral exemplarity: Just, brave, and caring. *Journal of personality and social psychology*, 86(4), p. 629.

Weisberg, M., 2013. *Simulation and similarity: Using models to understand the world.* New York: Oxford University Press.

Williams, P., 2009. *Mahāyāna Buddhism: The doctrinal foundations.* New York: Routledge.

Wittgenstein, L., 1958. *Philosophical Investigations I.* Translated by G. Anscombe. Oxford: Blackwell.

Appendix A

Intimacy/Affiliation

I help my friends and let them know I care. I spend some time with my closest friends. I try to be a close understanding friend. I strive to help those who are less fortunate than myself. I accept others as they are. I have more patience with my girlfriend. I have more patience with my boyfriend. I try to be a good listener. I give more than I expect to receive. I try to learn to express feelings of love toward my family. I try to be open about my positive feelings and be able to communicate them. I maintain and improve my relationship with my boyfriend. I maintain and improve my relationship with my girlfriend. I try to be friendly with others so they will like me. I seek to receive some form of physical affection. I meet new people through my present friends. I avoid being lonely. I want to be accepted in a social group. I try to keep in contact with long-distance friends so I will not lose them. I avoid being left out of conversations. I get enough attention and approval. I avoid confrontation with others.

Generativity

I try to be creative. I try to find ways to make the most of my creative abilities. I want to write and publish the book I have been thinking of for years. I want to contribute to my community. I want to provide help to my elderly and ill parent. I seek to be involved in improving my community. I try to help my kids with schoolwork. I try to help my kids play sports. I want to be a good father. I want to be a good role model for my children. I want to see that my children get the best education possible. I want to make a lasting contribution to the mission of my organization. I seek to feel useful to society. I try to make my life mean something. I try to make a lasting contribution to my profession.

Personal growth/health

I want to improve my health. I want to straighten out my values. I try to be more positive about myself. I want to be happier with my life. I want to develop a positive self-worth. I try to learn new skills and apply old ones. I try to avoid seeing the worst side of things. I try to avoid anxiety and other stressful emotions. I want to be strong while going through ups and downs. I try to be mentally aware of the things around me.

Spiritual transcendence

I want to deepen my relationship with God. I want to learn to tune into higher power throughout the day. I appreciate God's creation. I want to be filled with joy, peace and happiness from knowing God. I want to treat others fairly. I try to be more forgiving. I try to defend those who have been treated poorly. I try to deal with others with compassion and grace. I want to achieve union with the totality of existence. I want to become one with the cosmos. I try to immerse myself in nature and be a part of it. I try to approach life with mystery and awe.

Identity

I want to see beyond the myths and illusions that cause many of the problems within our society. I try to find that inner solace and higher mind. I seek new definitions for my life and what is good or desirable or realistic. I use charm and a smiling face to become friends with someone. I do my best in everything I undertake.

Achievement

I set high goals for myself and try to reach them. I realize my potential as far as a career is concerned. My goal is to achieve all that I can in as short a period as possible. I want to work towards higher athletic capabilities. I try to reach every career goal I set. I put my best effort into everything I do. I accomplish at least one main project a day. I strive to use my time productively. I think about my future career goals. I try to save money.

Power

I try to persuade others not to argue or fight with each other. I want to show that I am superior to others. I strive to get others interested in my work. I force men to be intimate in relationships. I strive to be the best when with a group of people. I want to be the dominant sibling in my family. I strive to get other people to do the things I want to do. I dress in an unconventional style. I organize social gatherings and bring people closer together. I argue my point to the end.

Self-presentation

I try to appear intelligent to others. I want to avoid appearing outrageous. I make myself physically attractive. I strive to impress others. I am concerned about my personal appearance at all times. I am self-confident in social situations when I am center stage. I am polite and well-mannered in the company of females. I act calm and cool when nervous. I always appear to be in a good mood. I try not to show negative emotions in public.

Independence

I do things without my friends every once in a while. I try to be myself and not do things to please others. I strive to not be a pushover with others. I am not dependent on my boyfriend. I set time aside to get my thoughts together. I am an individual. I want to become financially independent from my parents. I do not let other people take advantage of me. I want to stand up for myself and my beliefs. I do not depend on others to feel good about myself.

Human enhancement in contemporary society

5.

'Siri, tell me a joke': Is there laughter in a transhuman future?

Una Stroda

Independent Scholar

Abstract

The purpose of this paper is to explore the possibility of laughter, one of the basic aspects of human nature, continuing in the transhuman future. While laughter is universal, its expressions and content are culturally specific. Rituals, texts, and ideas from various traditions indicate the significance of laughter in human societies both as the source of life and a form of revelation. At the current stage of technological development, a sense of humor is one of the aspects lacking in artificial intelligence. Other challenges to the transhuman future of laughter include: the absence of a physical body to express laughter; the change of the locale of laughter from individual brain to collective mind; the lack of incongruity and surprise. The paper ends with a question: will transhuman laughter preserve any cultural meanings, physical expressions, and neural and psychological mechanisms of human laughter, will it be transformed, or cease altogether?

Keywords: Hospitality, Incongruity, Isaac, Joke, Laughter, Revelation, *Risus Paschalis,* Sarah

Introduction

Laughter is an aspect of human nature that is enchanting but also enigmatic. Laughter as a physical expression is universal and some elements of humor, for instance, laughing at incongruities and their resolutions, are appreciated by people of most cultures. However, although there are humor techniques, such as deception, disguise, exaggeration, irony or witty cynicism, that are perceived as consistently funny in different regions of the world, there is no universally funny joke. The content of laughter varies across the cultures because of the differences in cultural norms and practices. Culture, according to Hofstede, is the "collective programming of the mind which distinguishes the members of one group or category of people from another" (Hofstede,

1991, 4). Therefore, laughter, while present in all cultures, is conditioned by particular 'programming of the mind' or cultural norms and values that, to a large extent, are affected and influenced by religious beliefs.

For this reason, laughter as an aspect of human and possibly transhuman nature is considered here by predominantly working with religious perspectives. Furthermore, not only the content we laugh at is determined by norms, values and beliefs that are, in turn, shaped by religious traditions. There is also importance and quality attributed to laughter as such, articulated in a variety of religious contexts: ancient traditions, in particular, recognize laughter as an affirmation of life and celebrate its bodily expression. This point is of particular significance within the context of transhumanism that seeks to achieve the opposite: to transcend and overcome limitations of a human body. In addition, many religious traditions praise incongruous kind of laughter, considering it the sign of and a participation in divine revelation. Consequently, this article is divided into four parts: first, it describes and attempts to define laughter; second, it explores the value of laughter as life-giving and life-sustaining energy in religious settings that fashioned ancient and contemporary cultures; third, it examines the revelatory aspect of incongruous laughter; and fourth, it concludes with some thoughts on the future of laughter and evaluates specific cultural and theological challenges.

Siri, Apple's voice-activated assistant, is used here as an example of artificial intelligence that is increasingly becoming a part of our daily life. Because Siri is an AI assistant and plays part of the role that used to belong to human personal assistants, the implication is that Siri is in some of its functions equal to a human. There are times though when Siri attempts a joke but seems to be having problems with its sense of humor, indicating that AI lacks sensitivity, intuition and feelings that are distinctively human. Did 'collective programming of the mind' within particular cultural and religious contexts result in hardwiring sensibility for laughter into the human brain, thus making it essentially different from the transhuman mind? How important is laughter in various religious interpretations of what it means to be human? Should and will laughter continue into our transhuman future?

Theories of laughter

Despite the universal human ability to laugh, neural and psychological mechanisms of laughter are the subject of inconclusive speculations. Describing laughter itself is not difficult: laughter is a spontaneous and uncontrollable physical reaction, consisting typically of rhythmical audible contractions of the diaphragm and other parts of the respiratory system. It is also relatively easy to specify when laughter occurs: it can be a response to physical, psychological or intellectual stimulus. However, none of this explains why exactly human beings laugh. While a distinction can be made between "faked or felt" laughter (Ruch and Ekman, 2001, 27-28), most of laughing occurs spontaneously and unconsciously; we do not decide to do it, and very little is known about the brain mechanisms responsible for laughter. The evolutionary theory of laughter (MIT Staff, 2010) links it with the social brain hypothesis that explains the rapidly evolved brain as a result of living in larger groups: evolution of human language improved the fitness of individuals for life in groups, and laughter is simply a part of the same process. Another theory that associates laughter with emotions distinguishes between laughter as something we do and laughter that takes over us: there is "doing laughter" and being "done by laughter" or voluntary and involuntary laughter (Chafe, 2007, 66-68).

Some of the earliest philosophical attempts to explain and interpret laughter date back to Ancient Greece. Homer in his *Iliad* has gods laughing at weaknesses of both humans and other gods: Hephaistos, for example, is laughed at because he is unattractive and handicapped. This type of laughter is known as 'Homeric': both Plato and Aristotle considered it socially problematic. They encouraged laughing in moderation and even avoidance of all laughter, or at least jesting with good taste (Sanders, 1996, 106). Greek philosophers understood laughter as caused mainly by feelings of superiority of those who thought of themselves as being better, smarter, wealthier or more handsome than others. Superiority theory used to be a prevailing theory until the eighteenth century. Along with relief theory and incongruity theory, superiority theory is one of the most popular theories that seek to explain human laughter.

Relief theory originates with Sigmund Freud and situates laughter within the biological perspective: a person who is troubled by built-up nervous energy finds physical relief in laughter. Incongruity theory is more intellectual than the other two, and is the prevalent theory today. It sees laughter as an experience of the unexpected: someone or something behaves in such a way that contradicts one's expectations of how she or it should behave. Or, as Immanuel Kant in his often quoted phrase from *The Critique of Judgment* put it, laughter is "an affection arising from the sudden transformation of a strained expectation into nothing" (Kant, 2007, 161). Laughter results from the perception of disproportion, disharmony, or contrast between expectations and results. We laugh in reaction to a funny situation, a joke we hear, incongruity or paradox we perceive, or when we are surprised by a semantic discrepancy created by some dislocation of meaning.

These theories describe when and how we laugh, but they are insufficient to explain why the content of laughter varies between different groups of people and why there is no universally funny joke. A more recent theory of laughter, the benign violation theory, introduces the idea of cultural sensitivity into laughter research: according to this theory, laughter occurs when something feels wrong, unsettling or threatening, but simultaneously seems acceptable or safe (McGraw and Warner, 2014, 10). In other words, when a situation that is perceived as a violation of moral norms or social principles, turns out to be benign, people are relieved, feel amused and laugh. In addition to demonstrating the mechanism of laughter, the benign violation theory also supports the view that the mind is 'programmed'; it allows to explain why laughter is culturally determined and depends on moral norms and principles of particular groups of people. For example, a 2001 study of humor among Singaporean and American students (Nevo, 2001) found that Americans are more likely to tell jokes with sexual content while Singaporeans tend to focus on jokes containing aggression and violence. The researchers saw it as a reflection of broad cultural norms and traditions in which the participants of the study were brought up: American society is relatively liberal in matters of sexuality, while Singapore is more conservative and has a cultural trait that emphasizes strength and aggressiveness.

Laughter as a life-giving source in ancient beliefs and sense-making systems

Laughter, while universal, is shaped and influenced by cultural contexts and their differences: things people find funny depend on the 'collective programming of the mind'. Religious beliefs, cultural norms and traditions often determine what kind of laughter is acceptable. Such 'programming of the mind' takes place over extended periods of time in which religion has played a decisive role. Furthermore, the impact of religion is not limited to programming of moral norms and social values that influence laughter. In various traditions, laughter has acquired religious value and is envisioned as a life-giving and life-affirming source. Most profound examples come from the Ancient Near East, Israel, and classical antiquity.

Civilizations of the Ancient Near East considered laughter and also the opposite emotional expression, weeping, as the two most sacred emotional manifestations. They were actualized through rituals, ceremonies, and feasts remembering the death and celebrating the rebirth of various deities. During those annual feasts, the participants first wept over the death of a god and then rejoiced over his resurrection (Gilhus, 1997, 18ff). In Egypt this god was Osiris, in Babylonia it was Tammuz, in Canaan it was Baal, and there were similar cults in Syria, Phoenicia, Asia Minor, and Ancient Greece. Laughter that followed days of rituals of weeping over the dying and departing god celebrated the cosmic rebirth and marked a transition from mourning and fasting into communal celebration and festivities that included ritual jesting, clowning, obscenities, and laughter.

In Egypt, for example, the cult of Osiris who was impersonated by the Pharaoh developed during the period of the Old Kingdom. The Pyramid texts written during this period are part of the resurrection ritual for the Pharaoh and contain a ritual spell that alludes to the rite that included both lament and joyful celebration: "Sadness is over and laughter has come" (Allen, J.P., 2005, 118, 268). The cult of Osiris became even more important during the Middle Kingdom: the city of Abydos in Upper Egypt was established as a place of pilgrimage that gathered worshippers for the annual celebration of Osiris' death and coming back to life. While the exact procedures are largely unknown, it is believed that the public celebrations were combined with secret rituals or mysteries performed by the priesthood inside the temple (Mojsov, 2005, 52). Participation in passion plays that lasted eight days and concluded with "the jubilant rejoicing of the assembled crowds" (Mojsov, 2005, 52) was a way to identify with Osiris' suffering and death, at the same time sharing in his promise of eternal life.

Similar ideas of laughter as a sign of immortality and new life were present in rituals of other parts of the Ancient Near East. In Mesopotamia, the myth that told the story of Dumuzi and Inanna was re-enacted at the annual ritual of mourning. Once Marduk, the patron god of Babylon, became the head of the Babylonian pantheon, his death and the joy over his resurrection and his victory over Tiamat, the goddess of chaos and abyss, were celebrated at the annual Babylonian New Year's festival Akitu. In Canaan, Baal was the god of agriculture, rain, storm, and thunder. Tablet V AB of the Ugaritic texts, containing the Baal epic describes the resurrection of Baal that causes cosmic celebration for the return of fertility and abundance of life to the land, exciting the high god El: "The mightiest Baal is alive! Existent the Prince, Lord of Earth! ... The Kindly One El Benign is glad. His feet on the footstool he sets, and parts his jaws and laughs." As with other Ancient Near Eastern rituals of rebirth of the deity, laughter here marks the fulfillment, abundance, and the promise of new life.

A somewhat different character of laughter is reflected in yet another mythical story from Ancient Egypt, "The Contendings of Horus and Seth." The story describes the battles between Horus and Seth to determine who will succeed Osiris as king. Laughter occurs at the moment in the story when despair and depression set in among gods of Ennead who have not been able to reach an agreement, after hearing Horus and Seth argue for eighty years over who should inherit Osiris' kingdom (Morris, 2007, 198). The sun-god whose input was crucial in resolving the matter, had withdrawn to his tent, and at this moment his daughter appears: "After a long while, Hathor, Lady of the Southern Sycamore, came and stood before her father, the Universal Lord, and she exposed her private parts before his very eyes. Thereupon the great god laughed at her. Then he got up and sat down with the Great Ennead. He said to Horus and Seth, 'Speak for yourselves!'" (Morris, 2007, 199).

As this episode indicates, laughter can be not only an expression of joy after the deity awakens or comes back to life, but it can also be the stimulus for the deity to come out of the darkness, or, in case of the Egyptian sun-god, the tent. Interpreters of this skirt-lifting act or *anasyrma* point to the fact that in the Ancient Near East the notion of divine incest was not anomalous, and because Hathor was the goddess of physical beauty, her self-exhibition would have reminded her father of his potency in both political and sexual realms (Morris, 2007, 205). The laughter of the sun-god can also be explained as a reaction to the incongruity between the imagined and the actual reality: laughter was the result of the difference between how the sun-god expected his daughter to console him and what she in fact did. By means of obscene laughter that contradicts the norms of everyday life, a decisive turn of events that fostered a cosmic change and affirmed the value of life-giving laughter was achieved.

Another Egyptian myth where the sun-god literally creates the world through his laughter (Betz, 1992, 176-186) is a powerful vision of laughter as a life-giving source: "The god laughed seven times. When he laughed seven gods were born. When he laughed first, Light appeared and became god over the cosmos and fire." Each time the creator god laughs, a god is created. The seven new gods were Light, Water, Mind, Creative Force, Moira, Kairos, and Psyche. Once their creation was accomplished, the world came into existence and "he wept while laughing" (Betz, 185-186). Thus, laughter is pictured here as a powerful creative energy that could create life *ex nihilo.*

In the ancient rituals, death and renewal, suffering and joy, pain and laughter were parts of the same story repeated each year to affirm the core elements of life. Similarly, in Indigenous American spiritualities there is an underlined importance of laughter as a life-giving source within a context of two worlds: one is the "real world" and the other "not-so-real" (Laude, 2005, 97). This world is only a shadow of another, more real world, and it takes a fool, a clown, a jester, or a comedian to cast light onto the ambiguities of human existence, the weaknesses of humankind, and the danger of being lost in the chaos of life, thus missing the possibility of full reality (Laude, 2005, 97). The clown stands between the two worlds: he points to limitations and illusions of this world, and at the same time transcends it because he is a "manifestation of Being itself" (Zucker, 1969, 76). Moreover, the only way to overcome this 'not-so-real' reality is not to take it seriously, as suggested by Alves (1972): the advent of the full reality comes through the laughter of the clown. There is a similar idea in Christian theology: Paul introduces both himself and Christ as fools, and on several occasions suggests that Christians are "fools for Christ's sake" (1 Cor 4:10) or should become foolish (1 Cor 3:18), "for the wisdom of this world is foolishness before God" (1 Cor 3:19). Paul even talks about the foolishness of God: "Because the foolishness of God is wiser than men, and the weakness of God is stronger than men" (1 Cor 1:25).

The ancient ideas on laughter as a life-giving source were reaffirmed during the early Christian era in the theory of *Christus Victor.* This theory relates to Jewish apocalypticism popular before and around the beginnings of the Christian movement, the foundation of the Roman Empire, and the destruction of the temple. The apocalyptic thinkers suggested that God has relinquished control for the time being to the devil and its forces, but eventually God will intervene, overthrow the forces of evil, and establish the Kingdom on earth (Ehrman, 2006, 119). In early Christian thought Christ's death and resurrection were the means of defeating the powers of evil that held humankind in bondage of sin, death, and the devil (Aulén, 1969, 20). As the first followers of Christ were struggling to explain the mystery of Christ's death and resurrection, they understood it as God playing a joke on the devil: when the devil thought he has won because Christ died, God unexpectedly reversed the natural outcome to prove God is the winner. According to *Christus Victor* theory, the punchline of the divine joke is the event of the illogical and incongruous resurrection: Christ awakening from the dead is unexpected, unreasonable and surprising turn in the story, and as such resembles a joke.

The theme of Christ laughing at death re-emerged during the Middle Ages in the tradition of 'Easter laugh' or *risus paschalis.* The concept is based on the idea that at the original event, God played a joke on the devil by raising Christ from the dead; at Easter, Christians recalled this joke and laughed at it. The liturgical tradition of *risus paschalis* developed in fifteenth century Bavaria. At the end of the Easter Mass, the priest would leave the altar and come down to lead the congregation in what was called *risus paschalis.* He would tell funny stories and sing comical songs to induce laughter. The purpose of this ritual was to acknowledge in a tangible way through laughter as the most obvious expression of joy that Christ is alive and present among the people.

Incongruous laughter of revelation

Religious traditions of the Ancient Near East as well as Early Christianity found importance in laughter as a life-giving and life-sustaining source. Other traditions, such as Buddhism, Taoism, Judaism, or Christian Gnosticism, identify a different aspect of laughter as it relates to the divine: incongruous laughter as a sign of divine revelation. Notably, there are early Christian gnostic texts, most of them part of the Nag Hammadi library, containing references to laughter: *The Coptic Apocalypse of Peter*, *The Second Treatise of the Great Seth*, *The Apocryphon of John*, *The Hypostasis of the Archons*, and *The Gospel of Judas*. They express dualistic views of soul trapped inside the body and of a demiurge that creates the imperfect material universe. In these texts laughter is used to accomplish a specific doctrinal goal: to reinforce the Gnostic idea that the true followers of Christ possess gnosis that ensures salvation, and laughter is the sign of knowledge, insight and wisdom. For example, in *The Coptic Apocalypse of Peter*, Jesus is laughing at the moment of crucifixion. There are, in fact, two Christs: Peter is talking to Christ and realizing that there is another Christ being crucified at the same time. Jesus explains that the one "whom you saw on the tree, glad and laughing, this is the living Jesus." Those who are harming him do not realize that he is the divine being that cannot suffer or be killed; only his mortal shell can be. Once the body is killed, the inner self is released. Jesus laughs at those who know him 'in flesh' because in reality they do not know him at all.

In Gnosticism, the salvation of the soul was dependent on gnosis. Gnostics believed that the process of effectuating spiritual awakening and acquiring salvific knowledge is similar to solving a riddle or seeing a point of a joke (Gilhus, 75). Laughter at the moment of 'getting it' represented a passage from the hidden to the revealed, from ignorance to the divine truth. The conjunction of truth and laughter is not exclusive to Gnosticism; it is present in other traditions as well. For example, one of the translations of the *Tao Te Ching*, Chapter 41, reads:

> The wise student hears of the Tao and practices it diligently.
> The average student hears of the Tao and gives it thought now and again.
> The foolish student hears of the Tao and laughs aloud.
> If there were no laughter, the Tao would not be what it is
> (Lao Tzu, 1997, 80).

As this fragment indicates, laughter is an essential part of Tao. Without laughter, harmony with Tao is evasive, and the quest for truth is incomplete.

Similarly, laughter in the Buddhist tradition is the sign of insight and recognition. In Buddhism, there are at least two distinguishable types of laughter. One is a loud, open laughter of a kind that could "shake the very foundations of the world," and monks in Zen monasteries are said to be laughing more than in any other religious institution because laughter is considered a sign of sanity: laughter breaks up concepts, releases tensions, and teaches what cannot be taught in words (Hyers, 1974, 35-36). Another type of laughter manifest in Zen Buddhism is the smile of insight. There is a story of the Buddha who was gathered with his disciples, and instead of preaching a sermon, he simply held up the sandalwood flower before the assembly. None of those present understood the Buddha's meaning, except for one of the disciples, Kasyapa, who acknowledged his sudden intuition of Truth with a smile (Hyers, 1974, 23). This smile in the tradition of Zen is the signature of the sudden realization of the point or the Truth that can be neither taught nor learned; it is a Truth that only needs to be recognized and realized (Hyers, 1974, 137). Like the Gnostic laughter, Zen smile acknowledges the moment when truth is revealed: both imply the change of consciousness, recognition of incongruity and a paradigm shift towards greater understanding.

Incongruous laughter has a special meaning and value in Judaism. Jokes in Jewish tradition are popular because of their value to make a point and give insight to subjects beyond the joke itself (Berger, 1997, 89). In Judaism, the purpose of the joke is not to make a discourse funny but rather to enhance learning and understanding through a dialogue between the teacher and the audience. Laughter has always been valued as a way of insight into truth that cannot be taught in words. Humor, wit, and jokes are not simply a part of Jewish cultural distinctiveness; they are constituent elements of Judaism. For a rabbi, an essential skill is to be a storyteller, which in Jewish tradition is considered inseparable from joke-telling: the oral component of both implicates the idea of the discourse as an event or performance, and jokes are a way to engage the audience.

It is sometimes said that the Jewish predilection for joking begins with the incongruous agreement between absolute divine power and finite human beings: there is an insurmountable inequality of relationship between both (Wisse, 2001, 6). God and people enter into a covenant marked by such a great paradox that they have to laugh at it from the moment it is created. Incongruity between the finite and the infinite, the supernatural and the natural, the discrepancies between what human beings are and what they aspire to be, of what they can and cannot understand, and the paradox that God still has expectations for this covenant relationship to work is thought to be destined to result in laughter.

A special significance of laughter in Jewish tradition is indicated by the name of one of its patriarchs, Isaac, meaning 'laughter' or 'he laughs'. Before Isaac is born, both Abraham and Sarah laugh at God's promises several times. First, Abraham laughs at the moment of God telling him he will become "the father of a multitude of nations" (Gen 17:5). Abraham laughs again when he hears this promise: "Sarah your wife will bear you a son, and you shall call his name Isaac; and I will establish My covenant with him for an everlasting covenant for his descendants after him" (Gen 17:19, NIV). Without this promise, and the fulfillment of it, Jewish history would have started and ended with Abraham in the wilderness and no descendants, but Abraham laughs. Sarah, hearing she is going to bear a son, laughs too (Gen 18:12).

Postmodern feminist philosophy frequently turns to the image of Sarah because of the diversity of meanings that the image of a female laughing in a highly patriarchal Old Testament society implies. One of these interpretations links Sarah with the Ancient Near East emphasis on laughter, fertility, and creation: laughter, which is an opening of the body because we laugh with our mouths, can also be interpreted as a representation of fertility. Feminist philosophers speak of the visual similarity and the proximity of the meaning between the mouth and female genitals (Parvulescu, 16, 110). In this sense, laughter is the beginning of creation: there is laughter that comes out of Sarah's mouth, and the child named Isaac that comes out of Sarah's body. The theme of creation and laughter is also reminiscent of the Egyptian story of creation where gods are born out of the laughter of the creator-god. Even before Sarah laughs, there is the change of the names from Sarai and Abram to Sarah and Abraham (Genesis 17:5, 15). Their names are infused with laughter by virtue of the 'h' sound: every time their names are said out loud, with 'h' at the end, it sounds like a little laugh, and with every recounting of the story, the nations will be repeating the "laughter-inflicted" names of Sarah and Abraham (Parvulescu, 2010, 17).

Isaac is one of the patriarchs of Judaism, but he is the forefather of the Christian faith as well. In the New Testament, Paul asserts that Christ's followers are children of promise, like Isaac was (Galatians 4:28), and thus draws a parallel between Isaac and Christ. From the earliest Christian commentaries, the story of Isaac's sacrifice was seen as foretelling the death and self-sacrifice of Christ. Isaac, the Laugh, was spared from death; therefore, one can say that laughter overcame death. Christ, the new Isaac, conquers death and from that moment on, both Christ and his followers continue to laugh at death (O'Connell, 2002, 51) thus affirming the image of Laughter at the origins of humanity in Western culture and the belief that God's covenants are with the descendants of laughter.

The future of laughter

In many human cultures, laughter has been a life-sustaining source and a sign of divine revelation. This is changing because the human condition is being transformed by technologies that enhance human intellectual, physical, and psychological capacities. They also alter the way we function as social beings: our interpersonal relationships are affected by different enhancements, and technologies reshape the environment that surrounds us and situates us culturally, geographically, and historically. There are several challenges that transhumanism presents to laughter: for example, how will we define laughter if there is mostly a technological device and no human body to laugh? How would a collective mind, stored in the Cloud, react to humor? Will the ability of an augmented mind to foresee the future eliminate incongruity, surprise, and jokes? If there is laughter in the transhuman future, will it be culturally specific? Will laughter acquire a different value, or will it remain a sign of divine revelation?

One of the main challenges posed to laughter by the transhuman future is defining laughter itself. As described above, laughter is a physical reaction of our body. Is this going to change in the future? There are already examples in science fiction where personhood does not necessarily involve embodiment, such as the film *Her* (2013): this romantic science-fiction drama portrays a relationship between a man and a computer operating system personified through a female voice. The story of relationship free from the constraints of physical form uses a frame of transhumanism to reintroduce some age-old questions, such as what is it that makes love real? There are many ways in which the human body and its functions will potentially be altered. As a result, laughter might become a different physical expression, assume a separate, perhaps non-physical appearance, or might cease altogether. In addition, laughter as a physical reaction is also embedded in cultural heritage shaped by religious traditions, such as Christianity or Buddhism. Will the religious message be adjusted once a being lacking a physical body has trouble relating to the story of *risus paschalis* or the Laughing Buddha?

Second, structure and locale of laughter are already being adjusted and altered. Most of laughter originates in our mind as it notices incongruity, sends the impulse to our body to laugh, and gains insight from observing an issue from a new perspective. In the future, our brain can be enhanced by drugs, genetic engineering, or neuro-implants that can improve human intelligence and change our perception. Extended mind hypothesis suggests that our mind already is not confined to our brain: we delegate parts of it, such as memory or analyzing skills, to other devices. Once our mind is disseminated mostly outside our brain into other outlets, what part of our mind will send the impulse to laugh and where is it going to be located?

Ray Kurzweil, an American author, computer scientist, inventor, and futurist widely renowned for his work in the fields of AI and transhumanism, strongly believes that laughter will be part of the transhuman nature, arguing that artificial intelligence is already able to understand and to use puns, metaphors, riddles, and jokes (2012, 6): when the computer system Watson played *Jeopardy!* against the best two human players, it scored higher that the two of them combined and was able to master every area of human intellectual endeavor, including jokes and metaphors. Kurzweil also talks about advancement of *Homo sapiens* because of the evolution of knowledge made possible through technologies: the access to the cloud of computing resources contributes to the extension of mind, improvement of memory functions, and, ultimately, the extension of ourselves (2012, 279). Merging with intelligent technology will redefine biological limits of the human species: once the brain and other physical systems combine to form a mind, it might no longer be identifiable with individual contributors but rather becomes a collective mind, operating as a part of the Cloud. Assuming it is capable of laughter, how will this collective mind express its reaction to laughter?

Third, technological transformations of our social skills create uncertainty for the future of laughter in interpersonal communications. Not only our physical and intellectual abilities are being enhanced, but our social skills as well can be modified and improved with the assistance of artificial intelligence and augmented reality (Munkittrick, 2011). Specifically, our brains might someday be improved by implants that could enable us to analyze other persons' thoughts as well as predict their reactions and behaviors. While the enhancement techniques have not yet advanced to the stage where this is actually possible, there is already evidence that people become more trusting and cooperative when the level of the hormone oxytocin in their brains is increased: this eventually could lead to the development of drugs that increase our ability to empathize with others (Buchanan, 2017, 67) and to more and more advanced brain enhancement techniques and drugs. In the context of a social gathering, an enhanced brain would be able to analyze each person and determine their interests, personality type and social relevance to us. As a result, characters and reactions of others might become more predictable and less surprising for us. In terms of laughter, however, the ability to predict a social situation and its development makes it also less probable to be surprised by it or to encounter incongruity, such as a punchline of a joke, and laugh at it.

Technological transformations pertain not only to jokes but the concept of revelation as well. From religious perspective, revealing some form of truth through communication with a supernatural entity is essential for claiming the validity of truth. As illustrated by the examples from the tale of Kasyapa, some early Christian writings, and the tradition of Judaism, laughter can be a sign of recognition indicating the moment of comprehending truth that has been hidden before. Revelation implies that there is truth first unknown and then revealed through a surprising insight. The moment of revelation is often described as the point of becoming aware of truth, meaning, ultimate concern, personal relationship with the deity, or divine law. Revelation as the basic category of religious knowledge presupposes that there are profound insights into the meaning of life and the universe, received through an unexpected illumination from a disclosure of divine or sacred reality. If such occurrences of sacred disclosure become accessible to everyone through artificial assistance, would the theological concept of revelation as an exclusive event transforming human spirit lose its meaning and significance?

Fourth, transhumanism tests theological beliefs as they pertain to eschatology and the vision of the afterlife. Eschatological vision entails a new and qualitatively different stage of human existence, the age of perfection that is no longer affected by deficiencies, suffering, pain and death. If immortality becomes possible, what will then happen to eschatology? Will Zion, the eschatological place of joy where "our mouths were filled with laughter, our tongues with songs of joy" (Psalms 126:1-2), be relocated to this side of life because there is no longer any suffering caused by current imperfections of human body and pain inflicted by flawed environment? Joel Garreau, a journalist and an explorer of culture, values, history, and human nature, discusses DARPA (Defense Advanced Research Projects Agency) projects that are a part of their undertaking to create better soldiers (2005, 27). One research project, in particular, focuses on a 'pain vaccine' that would block intense pain by reducing inflammation and swelling. Other researchers are in the process of implementing photo-biomodulation that would accelerate wound healing, and other studies are concentrating on the natural chemical cascades in the human body that can stop bleeding if triggered by signals from the brain. Implications of these inventions will alter theological presuppositions, such as the Christian concept of incarnation: namely, the doctrine that Christ became flesh and fully participated in human suffering and pain so that humanity and all creation can be redeemed. As Christ's resurrection indicates, the greatest joy can be experienced only in the midst of the greatest suffering; once suffering and pain are minimized, what happens to joy? In the context of Indigenous American spirituality, Black Elk is quoted as saying, "truth comes into this world with two faces, it has both the laughing and the weeping face. Both are necessary" (Neihardt, J. G., 2014, 263). From the rituals of the Ancient Near East to the Christian story of the resurrection, laughter, while affirming life, has never been separated from the contexts of suffering, pain, and death. If human enhancements eliminate pain, the balance that reflects the fullness of humanity in its variety of physical and emotional expressions seems to be lost.

Fifth, the points mentioned above argue that human enhancements would have negative effects on laughter. However, laughter can also become an exceptional instrument to form hypothetical relationships with transhuman other. This theme of welcoming the completely other has been explored and elaborated by Jacques Derrida, who uses the subject of Sarah's laughter and interprets it within a context of welcome and hospitality. Laughter and smiling cannot be dissociated from hospitality: the welcome is hard to imagine without the smile or some sign of joy at the other (Derrida, 2002, 358). The welcome must include laughing or smiling, otherwise it is not a welcome at all.

Yet, contends Derrida, to be hospitable also means to be prepared to be unprepared for the other, to be surprised, to be overtaken, violated, raped and stolen, to let oneself "be swept by the coming of the wholly other, the absolutely unforeseeable stranger, the uninvited visitor, ... the absolute surprise" (Derrida, 2002, 161). There is no hospitality if the welcome is only of the known and the expected. Hospitality, as Sarah's case shows, is the experience of the impossible, the unknown, and the unexpected which she welcomes with laughter. Furthermore, there is not only the possibility of violence in the welcome of the other, in readiness to be overtaken by the other. There is also violence that precedes welcoming, an experience of "the Good that elects me before I welcome it," as Sarah was elected to show hospitality to the infinite (Derrida, 2002, 364). Sarah responds with laughter to the otherness that breaks in. Similarly, human laughter could be non-verbal, self-transcending, and, as the examples from various religious traditions indicate, incongruous and revelatory means of communication in meeting the otherness of the transhuman. From this perspective, laughter is decidedly an aspect of human nature that is fundamental to the transhuman future. The complete and utter otherness of the transhuman, of the unknown, of the stranger that ruptures in can only be encountered with Sarah's laughter of hospitality.

For Derrida, there is also the notion of "play" as a context in which the otherness of a different reality can be encountered (Derrida, 1978, 278-94). He derives 'play', one of his key concepts, from the semantics of language as an organized system where every element is ordered in relation to some invisible center or point of reference that, in fact, is not part of the system but is outside it. All the elements within the structure receive their meaning from other elements because they are relationally determined, and cannot have much freedom in receiving a different meaning apart from that which is assigned to them through being in relation to other elements. The center of this system can be supplemented by changing its name from God to a human person or idea, but it does not change the meaning of elements within the system. As an alternative, Derrida suggests releasing the structure of its defining center and bringing in the notion of play. Surrendering the structuring center would free the elements within the structure from being defined by their relations to other elements, and open the possibility of new meanings as well as expand the existing meanings. In terms of the future reality, Derrida's concept of play offers the possibility of new relationality between the human and the transhuman where laughter is not constrained to its humanity, its cultural limitations and definitions, and is instead open to re-definition, different understanding, and altered social expression and perception.

Conclusion

Transhumanism, on the one hand, is defined as a movement that affirms the possibility and desirability of fundamentally improving the human condition. On the other hand, transhumanism represents a transition between the human and the 'posthuman', a term describing a possible future being with basic capabilities so enhanced that they no longer can be called 'human' by our current standards. Thus, the posthuman and to a degree the transhuman as well, are the totally, completely, wholly other that can only be encountered, as Derrida suggests, with the smile and the laughter of welcome. However, the otherness of a transhuman other is still in the making and is unknown to us, therefore it is difficult to imagine the details of this encounter. In the posthuman future, both components of human laughter might be missing: the human body and the human mind. Having bodies could prove to be less advantageous than living as information patterns on vast computer networks, and minds that share memories and experiences directly could be found more efficient than limited individual human minds we have now. Thus, the totally other might have neither a body to laugh nor a cognitive technique to process and perceive laughter.

Moreover, the purpose of laughter as life-generating and life-affirming source can be supplanted by other means and technological achievements; laughter as an aspect of revelation might become theologically irrelevant; the timeframe of the eschatological vision of laughter and happiness as taking place in the fullness of times is going to be modified. Ultimately, will it even be possible for our transhuman selves to relate to Western culture with its strong Jewish and Christian tradition of laughter, of being surprised and amazed? This tradition starts with the astounding fact of creation itself and includes Isaac (literally 'laughter'); both Abraham and Isaac are considered the forefathers of faith and a reflection of God's image; there is early Christian tradition of God laughing at the devil in victory over death; there is a medieval tradition of remembering this laugh on Easter morning. Once our brain is so enhanced that we no longer are capable of laughing at jokes and incongruities of life, once we are no longer surprised or amazed by our existence, once we no longer can understand human jokes, will we still be the spiritual descendants of Isaac?

References

Allen, J. P. and P. Manuelian., eds., 2005. *The ancient Egyptian pyramid texts.* Atlanta: Society of Biblical Literature.

Alves, R., 1972. *Tomorrow's child: Imagination, creativity, and the rebirth of culture.* New York: Harper and Row.

Aulén, G., 1969. *Christus Victor: A historical study of the three main types of the idea of atonement.* Translated by Herber, A.G. SSM. New York: Macmillan.

Berger, P. L., 1997. *Redeeming laughter: The comic dimension of human experience.* New York and Berlin: Walter de Gruyter.

Betz, H. D., ed., 1992. *The Greek magical papyri in translation: Including the demotic spells.* 2nd ed. Chicago: University of Chicago Press.

Buchanan, A., 2017. *Better than human: The promise and perils of biomedical enhancement.* New York: Oxford University Press.

Chafe, W., 2007. *The importance of not being earnest: The feeling behind laughter and humor.* Amsterdam and Philadelphia: John Benjamins Publishing Company.

Derrida, J., 1978. Structure, Sign, and Play in the Discourse of the Human Sciences. In: A. Bass, ed. *Writing and Difference.* Chicago: The University of Chicago Press, pp. 278-293.

Derrida, J., 2002. Hostipitality. In: G. Anidjar, ed. *The acts of religion.* New York: Routledge, pp. 356-420.

Dormehl, L., 2017. *Thinking machines: The quest for artificial intelligence and where it's taking us next.* New York: A Tarcherperigee Book.

Ehrman, B. D., 2006. *The lost gospel of Judas Iscariot: A new look at betrayer and betrayed.* Oxford: Oxford University Press.

Garreau, J., 2005. *Radical evolution: The promise and peril of enhancing our minds, our bodies – and what it means to be human.* New York: Broadway Books.

Gilhus, I. S., 1997. *Laughing gods, weeping virgins. Laughter in the history of religions.* London and New York: Routledge.

Her, 2013. [Film] Directed by Spike Jonze. USA: Annapurna Pictures.

Hofstede, G., 1991. *Cultures and organizations: Software of the mind.* London: McGraw-Hill.

Hyers, C., 1974. *Zen and the comic spirit.* London: Rider and Company.

Kant, I., 2007. *The critique of judgment.* Oxford World's Classics. Translated by J.C. Meredith. Revised, edited and introduced by N. Walker. Oxford: Oxford University Press.

Kurzweil, R., 2012. *How to create a mind: The secret of human thought revealed.* New York: Viking.

Lao Tzu, 1997. *Tao Te Ching.* Translated by G. Feng and J. English. New York: Vintage Books.

Laude, P., 2005. *Divine play, sacred laughter, and spiritual understanding.* New York: Macmillan.

McGraw, P. and J. Warner., 2014. *The humor code: A global search for what makes things funny.* New York: Simon and Schuster Paperbacks.

MIT Staff., 2010. The evolutionary origin of laughter. *MIT Technology Review* [online] 29 October. Available at: <https://www.technologyreview.com/s/421480/the-evolutionary-origin-of-laughter> [Accessed 7 July 2018]

Mojsov, B., 2005. *Osiris: Death and afterlife of a god.* Oxford: Blackwell Publishing.

Moltmann, J., 1972. *The theology of play.* New York: Harper & Row.

Morris, E. F., 2007. Sacred and obscene laughter in the contendings of Horus and Seth, in Egyptian inversions of everyday life, and in the context of cultic competition. In: T. Schneider and K. Szpakowska, eds. *Egyptian stories: A British Egyptological tribute to Allan B. Lloyd on the occasion of his retirement.* Münster: Ugarit Verlag.

Munkittrick, K., 2011. When will we be transhuman? Seven conditions for attaining transhumanism. *Discover Magazine* [online]. Available at: <http://blogs.discovermagazine.com/sciencenotfiction/2011/07/16/when-will-we-be-transhuman-seven-conditions-for-attaining-transhumanism/#.worwwk6ngm8> [Accessed 14 February 2018].

Nag Hammadi Library. Available at: <http://gnosis.org/naghamm/apopet.html> [Accessed 2 November 2013].

Neihardt, J. G., 2014. *Black Elk speaks.* Lincoln: University of Nebraska Press.

Nevo, O., B. Nevo, and J.L. Yin., 2001. Singaporean humor: A cross-cultural, cross-gender comparison. *Journal of General Psychology,* 128(2), pp. 143-156.

O'Connell, M., 2002. Mockery, farce and *risus paschalis* in the York *Christ before Herod.* In: W. N. M Hüsken, K. Schoell, and L. Søndergaar, eds. *Farce and farcical elements.* Amsterdam: Editions Rodopi B.V.

Parvulescu, A., 2010. *Laughter: Notes on a passion.* Cambridge: The MIT Press.

Ruch, W. and P. Ekman., 2001. The expressive pattern of laughter. In: Kaszniak, A. ed., *Emotion, qualia and consciousness.* Tokyo: World Scientific.

Wisse, R., 2001. *Some serious thoughts about Jewish humor.* New York, Berlin: Leo Baeck Institute.

Zucker, W. M., 1969. The clown as the lord of disorder. In: C. Hyers, ed. *Holy Laughter: essays in religion in the comic perspective.* New York: Seabury Press, pp. 75-84.

6.

Making us better? Spirituality and enhancing athletes[1]

Tracy J. Trothen

Queen's University

Abstract

Elite sport is often where we see human enhancement technologies first and most visibly. What enhancements ought to be allowed in sports competitions is vigorously debated. This chapter makes an ethical case that if sport is one expression of the human quest for the sacred, this insight changes what it means to make athletes 'better'. The author theorizes that sport has a spiritual dimension for many followers which can be expressed in the hope that manifests in at least five locations in sport. Next, she distills four main approaches to the ethical assessment of sport enhancements and proposes that these four approaches would be reshaped if sport's spiritual dimension—as expressed in hope—is valued as much as is winning. The valuing of spirituality affects which enhancements we choose to accept and which we choose not to pursue. Tough questions are raised regarding values and who we desire to be.

Keywords: Human Enhancement, Spirituality, Sport, Ethics, Technology, Sacred, Interdependence

Introduction

The use of enhancements in sport is increasingly contentious. High profile athletes such as cyclist Lance Armstrong, sprinters Tyson Gay and Ben Johnson, baseball's Barry Bonds, and countless others have been thought to

---

[1] This essay and the keynote presentation on which it is based, builds upon and updates the foundations laid out in my book *Winning the Race? Religion, Hope, and Reshaping the Sport Enhancement Debate*. Macon, GA: Mercer University Press, 2015). I am grateful to Mercer University Press editor Marc Jolly for granting permission to publish this piece.

have unacceptable covert advantages in sports competitions. What makes an advantage unacceptable? When we participate in sports, usually we are improving ourselves and especially our physical condition. Sport is all about improvement, including the gaining of advantages over competitors.

Questions for us today go beyond which advantages are acceptable and which are not. We also need to go deeper and ask: Do sports enhancements indeed have a 'doubling effect' on athletic performances? In other words, do enhancements always make us even better than natural gifts, training, and effort alone? What does it mean to make us 'better'? If we believe that something makes us better, often we desire that something. What we desire reflects our values. So, what is it that we 'really, really want'?, as the popular breakout Spice Girls song asks. Answers to these questions influence how we assess the appropriateness and goodness of enhancing technologies and techniques.

The values that inform our desires and view of what makes us better are often unarticulated apart from scholarly inquiry; they are assumed and rarely explored on a self-conscious and reflective level. A question that I ask in this chapter is how the spiritual dimension of sport is morally relevant to enhancement use in sport. If sport has a spiritual or religious-like dimension, as many scholars have established (for example: Bellah, 1967; Novak, 1967; Albanese, 1982; Mathisen, 1992; Prebish, 1993; Price, 2006; Bain-Selbo, 2009; Austen, 2010; Cipriani, 2012; Ellis, 2014; Trothen, 2015; Bain-Selbo and Sapp, 2016), ought this dimension inform the assessment of enhancements? Is the spiritual dimension of sport of value to us? If the spiritual in sports is valuable, then we must ask how various enhancements affect the spiritual dimension of sports.

Using a Christian theological lens, in this chapter I ask how sport's spiritual dimension relates to the enhancement debate. First, I provide examples of sport enhancements. Next, I summarize my claim that sport's spiritual dimension is about hope.[2] More specifically, I describe five locations of hope in sport. In the second half of this chapter, I identify what I see as the four main approaches to sport enhancement ethics. I suggest how these four approaches might be influenced and reshaped if the locations of hope in sport—that is, sport's spiritual dimension--are valued as much as winning.

Enhancements in sport

Athletes have long used technology and science to improve their performances. Lightweight tennis racquets, biodynamic running shoes, training regimes,

---

[2] For a much more thorough development of this proposal please see Trothen (2015).

caffeine, banned substances such as EPO and anabolic steroids, and the list goes on. Now we are on the verge of new possibilities, including gene editing, robots, stem cell therapies, and moral bioenhancements.

Enhancements that may be attractive to athletes include the 'normal', visible add-ons, pharmaceuticals on banned substance lists, invisible performance enhancers that may or may not be on banned lists, 'natural' genetic anomalies, and interventions that fall in the murky zone between therapy and added extras. Genetic traits that we are born with can be contentious and viewed as 'unnatural' and too enhancing. Even though these traits are neither chosen nor intended by the athlete, the acceptability of these traits—such as hyperandrogenism in female athletes—may be heatedly debated. Some naturally enhancing traits are banned. And there are enhancing interventions that we choose but have become taken for granted and considered normal, such as vitamins and vaccinations. Techno-science innovations are resulting in more enhancement options for athletes, some of which are controversial and others not.

Performance-enhancing drugs such as anabolic steroids, HGH, and exogenous EPO are perhaps the best known enhancing substances and are banned by sport regulatory bodies. However, despite the awareness that these pharmaceuticals carry health risks ranging from reduced infertility to blood clots and major organ damage, and despite the recognition that cheating is wrong, there continues to be pressure on athletes to use banned substances. The recent widespread Russian doping scandal attests to coercion by coaches, and those with political and financial interests, on elite athletes to use banned substances.

Gene doping was added to the IOC's World Anti-Doping Agency's (WADA) banned list in 2003. As of January 1st, 2018, WADA's (2018) list also includes "the use of gene editing agents designed to alter genome sequences and/or the transcriptional or epigenetic regulation of gene expression." Possible gene therapies include therapies to prevent the breakdown of muscle tissue or to enhance red blood cell production (see, for example, Le Page, 2016). Samples from 2016 Olympic athletes were tested for added copies of a gene coding for EPO, a hormone that promotes red blood cell production, increasing athletes' endurance abilities. Although this gene therapy is still experimental, and meant for the treatment of renal diseases, it is very attractive to athletes who are desperate to stay in the game (Kirby, Moran, and Guerin, 2011) or to shave off a few extra seconds in pursuit of an elusive win. While there is general agreement among ethicists that enhancements should not be used unless, minimally, they are deemed medically safe, there is some debate regarding the acceptability of enhancements that might help create a more even playing field. Perhaps in the future some genetic modification technologies will be safe. Maybe these technologies could establish similar physiological capacities among elite athletes in given sports.

Is it fair and acceptable that genetic anomalies give many athletes a competitive edge through, for example, a greater lung capacity, greater ability to uptake more oxygen in the blood, longer limbs, or different hormonal levels. Usually, these 'natural' enhancements are accepted but sometimes they are not. For example, South African runner Caster Semenya was subjected to gender testing on the basis that she appeared too manly. Why are higher testosterone levels unacceptable in women's competitions but not in men's? Why are swimmer Michael Phelps' exceptional body proportions acceptable? Questions need to be asked around why some natural enhancements are acceptable and others not.

Functional Magnetic Resonance Imaging (fMRI) tests, which measure brain activity by detecting changes in blood flow, have shown that music, psychological techniques, and meditation or prayer affect the brain (Kobar et al., 2017; Schultz and Carron, 2013). Interestingly, in the 2007 New York marathon, there was an unsuccessful attempt to ban Ipods from the race since music was considered not only a safety risk but an unfair advantage (Macur, 2007). Are music, meditation, prayer, and even guided imagery exercises promoted by sports psychologists, forms of emotional doping?

Interventions that affect mood and cognition are becoming better understood and accessible. What about anti-depressants, since they change our mood? The question of where to draw the line between biomedical interventions that are therapeutic and meant to bring us up to a 'normal' status, and those interventions that go beyond therapy and an acceptable normal level is complex. Now we are talking about moral bioenhancements, as well. What if we can determine how to increase one's level of self- or team-interest, increase aggression, and decrease altruistic impulses? (Trothen, 2017) Surely such interventions could increase one's odds of winning.

Advances in equipment and training techniques are changing sports. Speedo's cutting edge 2008 Fastskin LZR led to the breaking of forty-three records at the 2009 international swimming federation (FINA) World Championship meet (O'Connor and Vozenilek, 2011). Since the polyurethane material reduced drag and muscle oscillation (Munroe, 2008), moving through the water became easier for swimmers. Concerns emerged regarding how these suits affected the basic test of swimming, and the fairness of the suits. On January 1st, 2010, FINA banned the use of these swimsuits in competitions (Partridge, 2011). The record wins stand but will, undoubtedly, be regarded as questionable since enhancing swimsuits were worn.

Artificial intelligence and genetic science are used to enhance training. For example, smart bats (see Zepp Staff, 2018) are being used in baseball training (to improve the batter's swing by increasing swing efficiency, speed, and power through feedback provided by a sensor inserted in the bat), and robots in football training. In the latter case, a team of undergraduate engineering students at Dartmouth College has developed the Mobile Virtual Player (MVP), a robot that can stand in to take concussion threatening hits during practice sessions. Although these robots are currently automated by remote control, work is being done to develop more autonomous robots that could participate in actual plays (Kastner, 2016). Nutrigenomics are informing optimal food intake for athletes. Predictive genomics have been used to help identify prospective star athletes even before they are old enough to participate in the sport (Kambouris et al., 2012).

It is not difficult to imagine the increased availability and selection of both external and internal enhancements for athletes. We have great sports sunglasses already—lightweight, gripping, and sleek. Why not sunglasses that add visual acuity and give us feedback about our performance? Or go a step further to retinal implants? This is not so far-fetched an idea given the recent development of bionic vision prosthetics, including retinal implants (Coffey, 2017). We now have externally worn fit bits. Aside from being less efficient, are these really any different from internal data chips or nanobots implanted in blood vessels on our brains? As opined in *Time Magazine* in September 2014, "the next product launch after the Apple Watch would logically be the iMplant" (Grossman and Vella, 2014).

To further complicate things, even reparative surgical interventions can be enhancing. In most instances, these therapies are clearly restorative (bringing someone back to what is widely accepted as normal after they have sustained an injury) but some enter, or are perceived to enter, the grey zone between restorative and enhancing. For example, Major League Baseball permits Tommy John surgery for pitchers who have torn their UCL in their throwing arm. Some think this surgery can make one's pitching even better than it was pre-injury. While medically this is disputed, and the need for the surgery is caused by poor biomechanics that can be corrected before the damage is done (Berra, 2007), the belief that it may have an enhancing effect has resulted in enthusiastic parents asking for their aspiring pitcher sons to receive the surgery (Longman, 2007).

Few people are asking what it is we really want out of sport. Is winning everything? Why is there so little conversation about how these innovations might interact with sport's spiritual dimension? These questions point to our values and desires. Of course, we want to win. But is there anything more we get out of sport? If there is more to sport than winning, then our conversations about enhancement use must go beyond how enhancements affect competition and winning.

The spiritual dimension of sport: hope

There is a significant body of interdisciplinary research that illustrates a strong relationship between sport and religion. Religious Studies scholar Joseph L. Price presents a strong argument that sport—especially baseball—satisfies Ninian Smart's seven dimensions of a religion (Smart, 1996).[3] More recently, Eric Bain-Selbo and D. Gregory Sapp make a similar case focusing on sport more broadly (2016). One of the earliest contemporary scholars to claim a relationship between sport and religion, the influential sociologist Robert Bellah, called sport a civil religion in the United States (1967). Others extend their analysis of implicit religions to include sport (Porter, 2009). Claims that sport is a natural religion (Novak, 1967), folk religion (Mathisen, 1992), cultural religion (Albanese, 1982), secular religion (Trothen, 2015), a popular religion (Price, 2006), a quasi-religion (Ellis, 2014) or simply a religion (Prebish, 1993) have been posited and examined.[4] A. Whitney Sanford understands whitewater kayaking as a source of transcendent experiences for some people (Sanford, 2007). Others choose to use the term spirituality more when describing this dimension of sport. Roberto Cipriani understands "sport as spirituality" (2012). Mike W. Austin sees sport as a potential part of spiritual formation (Austin, 2010).

Not all agree with these assessments. Some see sport as not related to religion or even as negatively related (Chandler, 1992; Higgs and Braswell, 2004; Hoffman, 2010). Some are very suspicious of characterizations of awe-inspiring experiences in sport as spiritual or religious (Higgs and Braswell, 2004, 219-224; Watson and Parker, 2013, 18). Many of these objections—but not all—seem to be at least partly precipitated by a concern that sport is displacing a transcendent God and so is heretical.

For or against, most scholars who write about religion, spirituality and sport, certainly see a relationship between sport and something religious-like or

---

[3] Smart's seven dimensions of a religion are: the practical and ritual dimension; the emotional and experiential dimension; the narrative or mythical dimension which includes sacred stories, saints, and prophets; the ethical dimension which includes rules and moral codes; the doctrinal and philosophical dimension; the social and institutional dimension which refers to organizational structures; and the material dimension which includes symbols and the preservation of sacred places.

[4] Nick Watson and Andrew Parker (2013) have authored an excellent 'mapping' of research on sport and Christianity.

spiritual. I build on this establishment of a relationship between sport, religion and notions of spirituality. Sport is not only about human enhancement in the technological sense. It is also and much more fundamentally about hope—human enhancement in the spiritual sense.

I suggest that sport centers on hope, and that the concept of hope overlaps with the concept of spirituality. After investigating thirty years of contemporary studies of hope from a spiritual theological perspective, pastoral theologian Pam McCarroll (2014) concluded that "Hope is the experience of the opening of horizons of meaning and participation in relationship to time, other human and non-human beings, and/or the transcendent" (48). Hope has been understood as an internal dispositional quality and, also, as transcendent. For McCarroll (2014), hope is not restricted to organized religion but is found in diverse human experiences arising in four domains: relationship with self, others, creation, and the transcendent. Meaning, transcendence, and present and future possibilities are key parts of hope, and key parts of sport. Hope can be associated with the pursuit of varied goals and outcomes—some specific and concrete and others more open-ended. Hope is subjective, contextual and necessitates the broadening of perspective as we discover meaning in diverse horizons. I propose that a key reason why sport draws so many followers and participants is that it *can* be a place where the sacred is discovered (Pargament, 2013) through the multifaceted presence of hope.

Sport can inspire hope in at least five different ways: winning, losing and anticipating; star athletes and 'my team'; perfect moments; embodied connections and possibilities of just communities; and flow states. I suggest that the sacred may be discovered in each of these locations of hope.

1. Winning, losing, and anticipation

The hope experienced in watching a team compete is not simply escapism. Real people, real dreams, real struggle and the overcoming of obstacles are involved in the journey to each competition. Fans and athletes share in the jubilation of winning and the despair of losing. But strangely, hope is not alive only in winning. Losing does not seem to crush hope in any lasting way–just ask a Toronto Maple Leaf's fan. Loyal Chicago Cubs fans persisted from 1908 to 2016 when they finally broke the infamous curse of the Billy Goat and won the World Series.

Part of the jubilation of a win is related to the identification many fans experience with their teams; winning can bring a special sort of hope that validates our loyalties and sometimes may even seem to validate ourselves. Psychologist Robert Cialdini and his research colleagues coined the phrase 'basking in reflected glory' (BIRG) to describe the phenomenon of fans identifying with their teams or players to the extent that the fans experience the glory of their team's win on a personal level as part of their own glory (1976). In an empirical study of 468 sports spectators and athletes, theologian Robert Ellis too found that "Winning...provides an affirmation of the self.... Winning seems close to salvation itself, it is a trip to the edge and back" (Ellis, 2012, 185). Taken to an extreme, winning can become the only way that one's identity is affirmed. Fortunately, Ellis' research showed that winning is not all that matters to most fans and players; respondents identified a "good quality game as important, a strong performance from one's team, [and] a good effort" as elements that also are meaningful and seem to stoke hope and loyalty (2014, 256).

The rhythm of seasonal competition is part of anticipation. Fans and players know that the next season or next contest is another opportunity and an important component of hope. Losing is not forever; neither is winning.

2. Star athletes and 'my team'

Identification with a team and their fan group makes one part of a community. Fans buy shirts with their favorite player's number. Games are watched faithfully and favorite players defended passionately. Star athletes are regarded by some fans, especially younger ones as found in Smith's 1976 Canadian study, as not only role models but moral exemplars worthy of worship (1976, cited in McCutcheon et al., 2004, 82).

Yet there is no rational reason to expect exemplary or even good moral behavior from someone known for athletic skill alone. Just as in formal religious organizations, power can be abused. Sport does not necessarily instill good character or virtue; athletes are not better people simply because they perform their sport well or even superbly.

There are many examples of athletes who have disappointed or even devastated fans when less than stellar aspects of their moral conduct have become public. The discovery of cyclist Lance Armstrong's use of performance-enhancing drugs meant the loss of his unprecedented seven Tour de France wins. Olympic runner Oscar Pistorius was charged and found guilty of murdering his girlfriend, Reeva Steenkamp. Track and Field's Marion Jones was stripped of three Olympic medals after being found guilty of steroid use. Complicating the picture is the reality that sometimes the athlete's moral compromise has been made for the sake of winning, or so it seems. One study found that the main reason athletes used banned substances was not to win but to stay in the game (Kirby, Moran and Guerin, 2011). As long as you are competing, the hope for the medal or the big victory stays alive. It is not only fans who cling doggedly to the hope that their sports star or team will succeed.

There are those athletes, such as many of the women in the Fédération Internationale de Football Association (FIFA) soccer world cup and Olympic hockey teams, who have committed themselves to inspiring young women and girls with the message that they can do anything and to take pride in themselves as gifted and amazing people with huge potential. As many feminist theologians have insisted, it is past time to redeem healthy pride amongst women and girls as a virtue consistent with Christian creation doctrine, which states that all people are created in the image of God with intrinsic dignity and worth. And there are also those athletes who disappoint us by behaving in less than admirable ways. Yet fans are determined to find worthy sports stars. As such, a relevant theological question is: given that we are all human, imperfect, and vulnerable to making mistakes, when is it morally good to continue to look up to an athlete after they make a mistake publicly?

Belonging to a sport community helps to shape one's identity, participate in meaningful activities with others, and to have inspiring performers or even perceived role models. These fan activities speak of hope: the opening of horizons of meaning and participation through a relationship with others.

3. Perfect moments

Witnessing a perfect athletic moment is a liminal experience in which the usually impossible happens. These perfect moments have been critiqued as often ego-centered and unworthy of associations with transcendence. But as Joe Price, writing on sport as a form of popular religion, reflects: "even when the pursuit of a perfect performance in sport becomes corrupted or distorted—when it moves toward selfish goals rather than the joy and disclosive possibilities of play itself—it still manifests a fundamental human desire for fulfillment" (Price, 2000, 211). This desire for fulfillment reflects a hope or longing for something more; we seek transcendence. There are moments in sport that edge into this transcendence and serve as a promise of something beyond.

These perfect athletic moments are mixed moments. As scholar of sociology and religion Michael Grimshaw (2002) suggests, they are instances of the sacred breaking into the profane—moments of the 'imperfect performer' bridging the supposed gap between the transcendent and immanent: Nadia Comaneci's perfect ten in the 1976 Olympic gymnastics, hockey player Paul Henderson's stunning goal in 1972 Canada-Russia series, Jayne Torvill and Christopher Dean's 1984 Olympic gold performance with 12 perfect 6s in pairs ice dancing. Somewhat different from Grimshaw's interpretation, I see these moments more as stark evidence that the sacred and profane, and immanent and transcendent are not binaries but are mixed in together, found in one another. Perfect athletic moments are fleeting but real, assuring us that what might seem humanly impossible is worth striving towards.

On the flip side, a relentless demand for a certain limited depiction of perfection is a distorted quest that sees healthy joy in doing the best one can devolve into a self-absorption that precludes all else and is never satisfied for long. A relentless demand for one version of perfection can also lead to excessive self-violence including extreme thinness, over-the-top risks in already high-risk sports such as snowboarding and whitewater rafting, and repeated over-training to injury. The moral assessment of violence is complicated, but at some point the harms caused by violence must outweigh the benefits.

This complicated relationship between religion, sport, and violence is exacerbated by the growing availability of risky enhancement means. As sports ethicist Sigmund Loland (2012) warns, the "logic of the record performance is that 'enough is never enough'" (162). Hope for record-breaking and winning alone reduces the meaning of perfection to a performance that conforms to certain expectations. And such perfect moments are never enough; they do not yield and manifest hope that opens horizons. Instead, meaning and possibilities are narrowed and do not allow one to remain inspired for long. Alternatively, when winning is not the only valued part of sport, and additional aspects such as a well-played game, diligence, fair play, and spirituality also are valued, perfect moments by imperfect humans can indeed inspire hope and open horizons of meaning and relationship.

4. Relational embodiment

Relational embodiment is the fourth location of hope in sport. This location of hope has two aspects: (i) Intra-relationality: integrated and embodied selfhood; and (ii) Inter-relationality: inclusion and the fostering of connection among diverse embodied persons.

(i) Sport has a strong potential to increase our awareness of the goodness of embodied personhood. Sport can visibly actualize and demonstrate the complex unity of body, spirit, and mind. Christian doctrines of creation, incarnation, and resurrection speak to the goodness of embodiment. At its best, sport helps us become more aware of our physicalities and the interconnection of all aspects of our beings. Without self-awareness, focus, determination, discipline, and optimism, it is impossible to do the best we can in sport. In short, sport can enhance our intra-relationality by pushing us to see more clearly how multiple aspects of being human interact—limiting and contributing to sports performance, enjoyment, and participation.

However, both sport and Christianity can fall short, sometimes dramatically, of espousing and manifesting the value of embodiment. We can push ourselves and each other too hard in the name of winning, or not hard enough. Instead of embracing and working with our complicated psyches, we may try to dismiss that aspect of our being or denigrate it.

When our various aspects of being human come together in sport, we become more whole and integrated. We move closer towards being fully human. In Christianity, full humanness and full divinity are inextricably linked in the incarnation of Christ. Becoming more fully human is always before us as a path to drawing closer to God.

(ii) Sport has the potential to not only help individuals to celebrate their embodied selves but also to unite a confluence of people from diverse racial, ethnic, and socio-economic backgrounds. For example, Nelson Mandela embraced the 1995 Rugby World Cup competition to unite South Africa in the wake of the dismantling of apartheid. In 1947 Jackie Robinson, as a star player and the first African American to play major league baseball in the modern era, contributed to the American Civil Rights Movement. Elite athletes with physical disabilities challenge assumptions about the meanings of disability. Hayley Wickenheiser and other women Olympic hockey players have helped to challenge gender stereotypes by following their dreams.

Sport both reflects and counters cultural norms; it can be both liberating and oppressive. Sport can reinforce discrimination based on numerous factors, including sex, gender, disability, race, ethnicity, age, and sexual orientation. Sport can also transcend and bridge these social prejudices. As theologian Jürgen Moltmann declared, Olympia is: "'a symbol of hope' if its character as protest, as alternative, and as the prelude to freedom is stressed, in its contrast to burdened everyday life in the economic, political and social world. This is a primal human longing" (Moltmann, 1989, 107).

5. Flow

Psychologist Mihalyi Csikszentmihalyi coined the term flow and has done groundbreaking work with his colleague Susan Jackson on flow states and athletes (1999). Aspects of a flow state include total absorption in the experience, the sense that all life is connected, and the loss of individual ego.

Flow experiences can provide the opportunity for one to see past the humanly constructed barriers that damage relationships. Or, the fan or athlete may enjoy the moment but fail to connect it to other aspects of life.

The sports fans' euphoric states have been variously called a wave, effervescence, flow, 'anonymous enthusiasm', or 'shining moments'. As philosophers Hubert Dreyfuss and Sean Kelly explain, "there are moments in sport—either in the playing of them or in the witnessing of them—during which something so overpowering happens that it wells up before you as a palpable presence and carries you along as on a powerful wave. At that moment there is no question of ironic distance from the event. That is the moment when the sacred shines" (2011, 194). In psychologist Kenneth I. Pargament's research, whatever is held sacred "act[s] like an emotional generator, stimulating feelings of awe, elevation, love, hope, and gratitude" (2013, 261-262). This fan energy can become manifested in benevolent or violent actions.

Flow can generate hope through the experiential awareness of the powerful connections between all life, including the transcendent, and through the sense of the extraordinary. However, intentional and communal reflection on the meanings of such awe-inspiring moments is necessary if these experiences are to be expressed in consonant actions. Perhaps intentional practice such as mindfulness meditation or team debriefing could assist participants in understanding flow experiences and the possible implications of these experiences.

If these locations of hope in sport are of value, then it is important to consider the moral relevance of hope to the sport enhancement discussion. In the next section, I identify four main approaches to sport enhancement ethics. I explore each approach by first providing a brief summary of the approach, next suggesting some critical questions, and then asking how the approach might be changed if sport's spiritual dimension of hope is explicitly valued. I suspect that normative values are assumed to the degree that we do not often identify, let alone question, the values that inform this ethics conversation. More deliberate examination and selection of a value base is necessary if we are to protect or even enhance the dimensions of sport that give sport meaning. If we begin with hope, the shape of the moral discourse changes.

Approaches to enhancement ethics and sport: What if hope is the starting point?

Implicit values permeate many aspects of life, technology included. European philosophers such as Habermas (1971), Marcuse (1964), and Foucault (1988) have established that technology is not value-neutral but is informed by and promotes the values of utility and efficiency. In the case of adding technology to sport, these values of utility and efficiency add to the normative emphasis on winning.

Theologically speaking, utility, efficiency, and winning reflect neither the sum value of being human nor the sum value of human competition in sport. Rather, there is intrinsic value in simply being, and there is intrinsic value in sport, win or lose. This does not mean that competition is unhealthy—part of the hope experienced in sport is a result of winning or trying to win. Winning only becomes problematic when it is everything or when its value eclipses the value of sports' intrinsic goods (MacIntyre, 2007). Similarly, it is not bad to be useful or efficient. But utility and efficiency become very problematic when a person's value is reduced to either. When enhancing technologies are used in sport, we need to be intentional about critically examining the values that might be amplified, expressed, or supported by these technologies.

The predominant approaches to the enhancement debate in sport tend to be decisionist, focusing on criteria for the acceptability of advantages or enhancements. This focus has led away from a critique of underlying values. These values need to be revisited and interrogated; otherwise we risk destroying what many cherish most about sport: its spiritual dimension of hope.

1. 'Individual choice' becomes 'relational autonomy and well-being'

Proponents of individual choice such as ethicists Andy Miah (2004) and Claudio Tamburrini and Torbjörn Tännsjö (2011) argue that athletes should have the freedom to choose whichever technologies they want so long as the enhancement is consistent with who they are, is deemed medically safe, and other athletes have the same freedom of choice. After all, we allow athletes to take significant health risks simply by playing their sports; why not allow them to choose enhancements that also carry risks? Detractors such as Robert L. Simon (2001 and 2004) and Sigmund Loland (2012) argue that while athletes do choose to accept risks that are intrinsically part of their sports, elite athletes often have their capacity to choose voluntarily whether to use banned enhancements unacceptably compromised. Coercion from owners, teammates, coaches, and others can leave the athlete faced with limited options: give up competing in their sport at a high level; follow the rules and likely lose to athletes who cheat; or bow to the pressure, accept the health risks and potential moral distress, and cheat.

A relational understanding of autonomy assumes that individual choice is complicated. Context is key to why we make the choices we do, and context includes systemic power imbalances and conflicting obligations. Choice is not entirely free and depends a lot on the perception of the power we have or do not have. This does not remove responsibility from athletes who decide to use enhancements–banned or not–but it extends responsibility beyond the individual.

Interestingly, even Miah who may well be the most extreme proponent of the individual choice criterion adds the caveat that the choice should be consistent with an individual athlete's 'authenticity' (2004), but seems to assume that access to one's authentic self, presumably including critical insight into the acquisition of values, is not too difficult. Uncovering authenticity that Miah requires a questioning of who I am and what it is I truly desire (i.e. knowing what I 'really, really want'). In theological terms, this includes asking in whose image am I created? As theologian Ron Cole-Turner warns, "technology turned on ourselves will teach us to see ourselves in a new way, as our own projects for improvement" (Cole-Turner, 2011, 7). Without a critical examination of the social processes that shape our choices (Sherwin, 2007), we accept assumptions that hinge on normative values such as acquisition, utility, efficiency, and winning.

If the spiritual dimension of sport is the starting point for an evaluation of sport enhancements based on choice, the individual choice approach would need to be reframed by a relational understanding of autonomy. A decision to use enhancement in sport is not just about that one athlete and his or her chances of winning. Studies such as Ellis' confirm that although winning is very important to athletes and fans it is not the only thing of value.

Theologically, when winning becomes the preeminent value in sport, sport risks becoming idolatrous. Winning becomes the sole image of the transcendent and of hope. When winning is the preeminent value, athletes are increasingly regarded as tools whose utility is all that matters. The enhancements we engineer and choose become oriented only toward winning and not to enhancing other values including hope. For example, the capacity to withstand more pain for the sake of beating one's competitor will be privileged over flow experiences, becoming more integrated beings, or teamwork. Genetic modification or editing of the 'DREAM gene' could cause pain sensations to be dampened or blocked (Miah, 2010). A protein called the Downstream Regulatory Element Antagonistic Modulator (DREAM) is associated with how we experience pain sensations and could theoretically be removed to allow us to endure more pain. So long as we normatively perceive vulnerability and dependence as undesirable and even as failings, choice will be influenced by avoidance of these human qualities. Theologically, vulnerability and interdependence are valuable aspects of being human that emphasize our communal character and mutual need of each other. From a theological perspective that values interdependence, instead of blocking our capacity to feel pain, we may be very interested in the potential of digital aids, such as some Fitbits, that can help us connect with other sports participants for mutual encouragement.

By renaming the 'individual choice' approach to sport enhancement ethics as 'relational autonomy and well-being', I intend to indicate two shifts that would occur if we begin the ethics conversation with an assumption of the existence and value of sport's spiritual dimension: 1) a broader understanding of autonomy that includes rights but extends beyond individual rights to include individual and communal responsibility; and 2) a broader understanding of well-being that includes avoidance of physical harm, and expands to include doing good. Our capacity to identify our desires is molded to a large extent by a culture increasingly driven by technology and acquisition. We also need to explore what additional rights and responsibilities are important to human flourishing. In so doing, the possibility of a more fulfilling life increases. The reduction of autonomy to individual choice can mean the neglect of relevant systemic issues such as the distribution of power, the impact of an individual's choice on others, and the social processes that inform desire and choice.

We need to attend to these issues communally if we are to make more fully informed and authentic choices. Otherwise, the uncritical choice to use enhancements may cause broader harms, including damage to sport's spiritual dimension. Authenticity is about more than individual desire. And fulfillment and satisfaction are not achieved by winning alone.

2. Fairness: Sameness or diversity?

Extreme proponents of the fairness approach see choice and the flattening of difference as desirable (Tamburrini, 2007, 234). Although at the time of the writing of this chapter, there have yet to be any confirmed cases of athletes genetically enhancing their bodies [gene doping would be challenging to detect apart from the developing athlete biological passport, and CRISPR (clustered regularly interspaced short palindromic repeats) will make such therapies easier to administer and even harder to detect], there are rumours of the use of gene therapy to block the breakdown of muscle tissue, and for other purposes that could be useful to athletes wanting to improve their performance.

For example, Repoxygen was the tradename of a gene therapy under preclinical development for the treatment of anemia but in 2003 was abandoned likely due to anticipated minimal profit. In 2006, German track coach Thomas Springstein was found guilty of trying to procure Repoxygen for his athletes (Aschwandan, 2012; Friedmann, Rabin and Frankel, 2013, 647). Repoxygen or a similar gene therapy could increase the oxygen-carrying capacity of red blood cells, and arguably even-the-playing-field by minimizing the differences among athletes' endurance levels; every competitor could have the edge possessed by Finnish athlete Eero Mäntyranta who, in the 1964 Winter Olympics, won two gold medals in cross-country skiing. Mäntyranta had a rare mutation in the gene coding for his EPO receptor that resulted in a 25–50 percent increase in the oxygen-carrying capacity of his red blood cells.

Less extreme proponents of the fairness approach want to minimize only the differences that are considered *unfair* advantages, not differences that are considered merely fortunate. But how we decide whether differences are unfair or simply lucky is influenced by our values and prejudices. For example, Caster Semenya had to undergo gender testing to decide if she an unfair hormonal advantage since she looked too manly, but swimmer Michael Phelps' unusual 'wingspan' was seen as lucky. Sports regulating bodies such as WADA have a difficult task in determining which advantages are acceptable and which give an unfair advantage. In making these decisions, it is crucial to be self-critically aware of which bodies are more socially valued and how this valuing might influence decisions about advantaged of genetic happenstance. Are we seeking fairness or the advancement of a normative body type?

The minimizing of difference has at least two problematic implications for elite sport. First, as enhancement options multiply, sport likely will become more of a performance of 'competing bodies' and less of a 'contest between persons' as ethicist Simon puts it (2001). Some have argued that the widening of the gap between elite athletes and everyone else is desirable; that is why people watch—for entertainment (Tamburrini and Tännsjö, 2011). Others disagree, pointing to the identification of spectators with the athletes (Ellis, 2012). The second implication goes hand in hand with the first: there will be a further debunking of the illusion that sport is a meritocracy and so competitively available to anyone. This loss of the illusion of meritocracy, and therefore possibility, means a loss of a source of popular hope and meaning. The flattening of body differences will not add meaning to sports competition; such competition is built on human diversity.

There are at least three faulty assumptions with the argument that fairness will be increased by flattening differences: 1) it is assumed that by replicating physical advantages, athletes will have the same capacities, but this claim does not account for the gestalt of being human–we are more than the sum of our parts; 2) it is assumed that we know many of these existing advantages and soon will be able to identify all advantages (we will learn more, but again, advantages are complex and cannot be reduced to any one gene or even group of genes); and 3) it is assumed that athletic competition is ideally performed between people who are the same physically.

Diversity is needed for athletic competition and for the generation of hope: without the sense of the imperfect performer sometimes having perfect moments, there would be no bridge that allows us to believe that anything is possible. Sport hinges on the diversity of athletes. No one is the same in elite sports, but all possess exceptional qualities.

Fairness is an important criterion for evaluating an enhancing technology, but the meaning of fairness needs attention. Fairness is not sameness if we value the diversity of people as created in the image of God. Fairness ought to involve social justice if relational embodiment and the hope that goes with this is valued. If hope is our starting point, uneven power and resources are what need flattening, not embodiment diversity.

3. Therapy-Restoration-Enhancement

This is a limited but useful approach to sport enhancement ethics. This approach relies on the notion of 'normal' as a dividing zone between therapy and enhancement. If an intervention is judged to fall within the therapeutic spectrum, it is acceptable. If it is enhancing, it is usually not permitted unless it is deemed close to normal or inconsequential. The middle zone of interventions that are neither purely therapeutic nor clearly enhancing is complex, especially since some interventions become normalized and accepted over time, and others currently in use prove risky or undesirable. But generalities are useful even though there are exceptions.

Performance-enhancing drugs include anabolic steroids at one end of the continuum due to their clear health risks, measurable effect on performance, and banned status. Typically, they are used for one purpose: to gain a covert competitive edge. At the other end of the continuum are therapeutic interventions such as ultrasound treatment for a soft tissue injury, or most surgeries to repair torn ligaments.

Somewhere in the middle of the continuum is the use of hypobaric chambers since some argue that these tents can be used to both help one train more fairly and to to enhance an athlete's aerobic capacity above what is considered normal. Also in the mid-zone are interventions that arguably restore one's capacities to a level that is considered normal. But what is considered normal shifts according to context. In elite sport, normal physical capacities are higher than in the general population. Is it restorative, and therefore acceptable, to improve a Major League Baseball player's eyesight to greater than 20/20 since most MLB players have greater than 20/20 eyesight? The limit of this approach will continue to be the questions of when exactly does therapeutic become enhancing, and at what point do enhancing interventions go beyond restoration to generate unfair advantages.

A problem with the enhancement end of the continuum is that some enhancing interventions do more good than harm; not all interventions that bring us beyond some notions of normal are bad. There may be many advantages to the enhancement end of the continuum depending on what it is we decide to enhance, and whether these enhancements are in line with values that we wish to propagate.

Again, which values we select depends upon the lens through which we view the world. From some theological perspectives (there is much diversity, of course, within faith traditions), Christianity underscores the goodness of embodiment in the doctrines of creation, incarnation, and resurrection. Further, biblical stories about healing demonstrate the goodness of using our abilities to promote healing. The question of what healing means is connected to the question of what enhancing means. Theologically, we are called to discern our way back towards a prelapsarian state, but we cannot do this apart from God's grace and redemption.

Motive then becomes important in adjudicating which enhancing interventions are congruent with an openness to and desire for spiritual qualities. For example, if hope is found in relational embodiment in sport, then the preservation of this hope requires attention to the well-being of others beyond any one athlete. If my pursuit of an enhancement fails to promote, or even detracts from the well-being of others, then it is not a justifiable choice.

For now, it seems that utility and efficiency associated with winning are driving the interpretation of enhancing. If hope in sport is our starting point, to assess interventions in the mid-zone of this continuum, we must ask whether the intervention promotes or damages the five locations of hope. Winning alone is not a sufficient criterion to justify an intervention that falls in the mid-zone.

4. The meaning of sport: hope as core to sport

Increasing attention is being given to the meaning of sport for assessing enhancements. Approaches to the enhancement debate that focus on values are distinct in that they tend to be more virtue ethics-based and less decisionist than the other approaches. I will touch on two main issues that I include in this fourth approach to sport enhancement ethics: 1. The basic tests of sport and 2. The broader question of what is valuable in sport. The argument offered by a few scholars is that the degree to which the basic tests of sport and/or the values and meaning of that sport are affected ought to be the main criteria for evaluating enhancements.

1. The basic test of a sport defines the structure of that sport and the rules of competition. For example, FINA judged that polyurethane super swimsuits made the basic test of swimming too easy and were therefore banned in competitions. Another example takes us back to

the case of Oscar Pistorius. Critics charged that what he did was not running since he did not use two natural legs (Burkett, McNamee, and Potthast, 2011). One can imagine additional possible ways that enhancement technology might unduly affect the basic test of a sport. What about the use of a skate in hockey that heats itself to more effectively melt ice and thus increase skating speed? Would that invalidate the basic test of skating well? Or what if we could use genetic modification technologies to increase a batter's strength, and visual acuity, so as to greatly improve their chances of hitting home runs? Would that not invalidate one of baseball's basic tests? The overall concern is that continued development and use of enhancements will change the nature of sports to the point that each sport is no longer recognizable. The basic test of a sport will be too easy (as in the case of polyurethane swimsuits) or inappropriate (as some argue regarding Pistorius). Further, the hope, awe, and mystery will become dampened as sport becomes more about extreme entertainment and less about "incarnate spirits achieving wonders" (Carmody, 1986, 177).

2. The meaning of sport includes internal goods, which are "intrinsic satisfactions that come from meaningful engagement in them" (Morgan, 2006, 249). Internal goods are not utility-oriented, are accessible to anyone involved, and benefit the "whole community who participates in the practice" (MacIntyre, 2007, 190-191). Both internal and external goods (such as winning, financial awards and contracts, social status and fame) are important to sports and, as ethicist Alasdair MacIntyre argues, the pursuit of neither is necessarily bad unless the acquisition of external goods such as winning supersedes the value placed on internal goods (2007, 190-191). While there are some shared internal goods among all sports such as fair play, doing one's best, discipline, and teamwork, there are some internal goods that are more important to some sports than others. For example, excellence of performance is an internal good for all elite sport but is defined in a particular way for each sport. Excellence in hockey generally requires reliable passing plays and the ability to communicate almost intuitively with one's line-mates. This orchestration depends on both the individual player and communication and connection between the players. Excellence in bodybuilding is more reliant on individual work characterized by discipline, focus, drive, and technique. Internal and external goods of sport, and values that are not constrained to sport, all contribute to the meaning of sport. I think one of the most important values and

> internal goods that shape the meaning of sport is spirituality as manifested in hope.

Spirituality in sport centers on the capacity of sport to inspire hope and, as part of hope, a sense of meaning. It is hope that spurs us on to overcome obstacles and to search for fulfillment and meaning. Hope is rarely identified in sports literature as a normative good, outside of sources specific to the relationship between sport and religion. Even in these sources, the religious and spiritual dimension of sport is sometimes seen as more suspect than not. If hope is a core value to sport, then enhancements must be considered primarily in view of how they might affect hope as it is experienced in sport. This would shift the ethics discussion.

If sport's spiritual dimension is highly valued, then the locations of hope must be prioritized in the assessment of enhancement use. If an enhancement would negatively affect the inspiration generated by star athletes and cherished teams, or dampen the breathtaking awe of a perfect moment, or further reduce elite athletes to instruments, or limit sport's potential for serving as a meeting ground for diverse people, or prevent the possibility of flow experiences, then they are not enhancing enhancements. For example, if inter-relationality as a location of hope in sport is valued, then oxytocin will not be used as a potential moral bioenhancement in sports. Moral enhancement is garnering attention as a potential way to promote normative virtues and values and theoretically safeguard the planet and humans from mass destruction. While oxytocin increases altruistic behavior and empathy, it only does so towards in-group members (Jones, 2013, 190; Persson and Savulescu, 2015, 338). As a result, the usage of oxytocin may increase enmity between international teams, particularly in the Olympics, by binding us closer to kin and, therefore, possibly causing us to become even less concerned with the welfare of others. However, if the external good of winning is prioritized over internal goods associated with sport, especially hope, then oxytocin may become very desirable if it increases loyalty to one's team at the expense of the competition. When we begin with the conviction that inter-relationality is a valued location of hope in sport, we will be more interested in enhancing sport through greater efforts to make improving technologies equitable and much more globally accessible, than we are with winning at any cost.

This last approach to sport enhancement ethics may hold the greatest possibility for introducing hope in sport as morally relevant to the enhancement issue. If hope is recognized and seen as core to the meaning of sport, enhancements that might negatively affect hope would be neither acceptable nor desirable. As an addendum, it may even be argued not only that hope is a preeminent value germane to sport's meaning, but that hope also protects the integrity of sports' basic tests. As we have seen, much of the hope in sport relies on the ability of people to identify with athletes as vulnerable human beings. If we accept enhancement technologies that lay waste to the basic tests of sport, then much of the hope in sport will also be lost.

Back to the beginning: what do we value?

This chapter touches the surface of very complex ethical questions concerning sport enhancement ethics and hope. My intent has not been to offer a thorough exploration of hope in sport and the ethical approaches to enhancement but to demonstrate that hope is a neglected and morally relevant domain to the sport enhancement conversation.

Hope is intertwined in the meaning of sport. Sport is meaningful in large part because of the hope that is embedded in our competitions and games. Sport cannot be enhanced without preserving and possibly amplifying hope. The meaning of 'enhance' is key. What counts as enhancing—or making better—depends on what one thinks is most valuable about being human. The most important theological dictum relating to the sport enhancement debate may not be the Promethean precaution to avoid thinking of ourselves as gods, but the proactionary exhortation to become as fully human as possible. As with traditional organized religions, sport can distort the very things that make it shine. Even with these distortions, sport inspires hope and brings out some of the best qualities we have. If hope is an internal good of sport, how might this insight reshape the sport enhancement debate? The question of how an enhancing technology might affect spiritual experiences and especially hope in sport would become vital to the discussion. The identification of the spiritual dimension of sport as a core value would also shift attention to the question of how to understand the functions of hope in sport and how to enhance this hope. Techno-science enhancements need to be assessed first and foremost with the question of how these enhancements will affect hope. And then . . . We all win.

References

Albanese, C., 1982. *America: Religions and religion.* Belmont, CA: Wadsworth.

Aschwandan, C., 2012. The future of cheating in sports. *The Smithsonian Magazine* [online]. Available at: <http://www.smithsonianmag.com/science-nature/The-Future-of-Cheating-in-Sports-160285295.html> [Accessed 24 January 2018].

Austin, M. W., 2010. Sports as exercises in spiritual formation. *Journal of Spiritual Formation & Soul Care,* 3(1), pp. 66–78.

Bain-Selbo, E., 2009. *Game day and god: Football, faith, and politics in the American South.* Macon, GA: Mercer University Press.

Bain-Selbo, E. and D.G. Sapp., 2016. *Understanding sport as a religious phenomenon: An introduction.* New York: Bloomsbury Academic.

Bellah, R. N., 1967. Civil religion in America. In: W.G. Mcloughlin and R.N. Bellah, eds. *Religion in America.* Boston: Houghton Mifflin, pp. 3-23.

Berra, L., 2012. Force of habit. *ESPN* [online]. 23 March. Available at: <http://www.espn.com/mlb/story/_/id/7712916/tommy-john-surgery-keeps-pitchers-game-address-underlying-biomechanical-flaw-espn-magazine> [Accessed 24 January 2018].

Burkett, B., M. Mcnamee, and W. Potthast., 2011. Shifting boundaries in sports technology and disability: Equal rights or unfair advantage in the case of Oscar Pistorius? *Disability & Society*, 26(5), pp. 643–654.

Carmody, D. L., 1986. Big-time spectator sports: A feminist Christian perspective. *New Catholic World*, July/August, pp. 173-177.

Chandler, J. M., 1992. Sport is not a religion. In: S. J. Hoffman, ed. *Sport and religion.* Champaign, IL: Human Kinetics Books, pp. 55-61.

Cialdini, R.B., R.J.Borden, A. Thorne, M.R. Walker, S. Freeman, and L.R. Sloan., 1976. Basking in reflected glory: Three football (field) studies. *Journal of Personality and Social Psychology*, 34, pp. 366-375.

Cipriani, R., 2012. Sport as (spi)rituality. *Implicit Religion*, 15(2), pp. 139-51.

Coffey, V. C., 2017. Vision accomplished: the bionic eye. The optical society of America, Inc. [online]. April. Available at: <https://www.osa-opn.org/home/articles/volume_28/april_2017/features/vision_accomplished_the_bionic_eye/> [Accessed 24 January 2018]

Cole-Turner, R., 2011. Introduction: the transhumanist challenge. In: R. Cole-Turner, ed. *Transhumanism and transcendence: Christian hope in an age of technological enhancement.* Washington, DC: Georgetown University Press, pp. 1-18.

Dreyfuss, H. and S.D. Kelly., 2011. *All things shining: Reading the Western classics to find meaning in a secular age.* New York: Free Press.

Ellis, R., 2012. The meaning of sport: An empirical study into the significance attached to sporting participation and spectating in the UK and US. *Practical Theology* 5(2), pp. 169-188.

Ellis, R., 2014. *The games people play: Theology, religion, and sport.* Eugene, Oregon: Wipf & Stock.

Foucault, M., 1988. *Technologies of the self.* Boston: University of Massachusetts Press.

Friedmann, T., O. Rabin, and M.S. Frankel., 2013. Gene doping and sport. *Science* 327, pp. 647–648.

Grimshaw, M., 2002. I can't believe my eyes: The religious ascetics of sport as post-modern salvific moments. *Implicit Religion*, 3(2), pp. 87–99.

Habermas, J., 1971. *Knowledge and human interest.* Boston: Beacon Press.

Higgs, R. J. and M.C. Braswell., 2004. *An unholy alliance: The sacred and modern sports.* Macon, GA: Mercer University Press.

Hoffman, S. J., 2010. *Good game: Christianity and the culture of sports.* Waco, TX: Baylor University Press.

Grossman, L. and M. Vella., 2014. How Apple is invading our bodies. *Time Magazine* [online]. Available at: <http://time.com/3318655/apple-watch-2/> [Accessed 24 January 2018].

Jones, D. G., 2013. Moral enhancement as a technological imperative. *Perspectives on Science and Christian Faith*, 65, pp. 187–95.

Jackson, S. and M. Csikszentmihalyi., 1999. *Flow in sports: The keys to optimal experiences and performance.* Champaign, IL: Human Kinetics.

Kambouris, M., F. Ntalouka, G. Ziogas, and N. Maffulli., 2012. Predictive genomics DNA profiling for athletic performance. *Recent Patents on DNA Gene Sequences*, 6(3), pp. 229-39.

Kastner, E., 2016. How a robot football player will prevent concussions: During practice, the MVP robot can stand in for American football players and take the tackles. *IEEE Spectrum* [online]. Available at: <https://spectrum.ieee.org/robotics/humanoids/how-a-robot-football-player-will-prevent-concussions> [Accessed 24 January 2018].

Kirby, K., A. Moran, and S. Guerin., 2011. A qualitative analysis of the experiences of elite athletes who have admitted to doping for performance enhancement. *International Journal of Sport Policy and Politics*, 3(2), pp. 205–24.

Kobar, S. E. et al., 2017. Ability to gain control over one's own brain activity and its relation to spiritual practice: a multimodal imaging study in frontiers. *Human Neuroscience*, 11. DOI 10.3389/fnhum.2017.00271

Le Page, M., 2016. Gene doping in sport could make the Olympics fairer and safer. *New Scientist* [online]. 5 August. Available at: < https://www.newscientist.com/article/2100181-gene-doping-in-sport-could-make-the-olympics-fairer-and-safer/> [Accessed 24 January 2018].

Loland, S., 2012. A well balanced life based on "the joy of effort": Olympic hype or a meaningful ideal? *Sport, Ethics and Philosophy*, 6(2), pp. 155–65.

Longman, J., 2007. Fit young pitchers see elbow repair as cure-all. *New York Times* [online]. 20 July. Available at: <http://newyorktimes.com/2007/07/20/sports/baseball/20surgery.html> [Accessed 24 January 2018].

Macintyre, A., 2007. *After virtue: A study in moral theory.* 3rd ed. Notre Dame, IN: University of Notre Dame Press.

Macur, J., 2007. Rule jostles runners who race to their own tune. *New York Times* [online]. Available at: <http://www.nytimes.com/2007/11/01/sports/othersports/01marathon.html> [Accessed 24 January 2018].

Marcuse, H., 1964. *One dimensional man: Studies in the ideology of advanced industrial society.* Boston: Beacon Press.

Mathisen, J., 1992. From civil religion to folk religion: The case of American sport. In: S. J. Hoffman, ed. 1992. *Sport and Religion.* Champaign, IL: Human Kinetics Books. pp. 17-33.

Mccarroll, P. R., 2014. *The end of hope: The beginning: narratives of hope in the face of death and trauma.* Minneapolis: Fortress Press.

Mccutcheon, L. E., J. Maltby, D.D. Ashe, and J. Houran., 2004. *Celebrity worshippers: Inside the minds of stargazers.* Baltimore, Maryland: PublishAmerica.

Miah, A., 2010. The DREAM gene for the posthuman athlete: Reducing exercise-induced pain sensations using gene transfer. In: R.R. Sands and L.R. Sands, eds. 2010. *The anthropology of sport and human movement: A biocultural perspective.* Plymouth, UK: Lexington Books. pp. 327-341.

Miah, A., 2004. *Genetically modified athletes: Biomedical ethics, gene doping, and sport.* New York: Routledge.

Moltmann, J., 1989. Olympia between politics and religion. In: G. Baum and J. Coleman, eds. 1989. *Sport.* Edinburgh, UK: T&T Clark. pp. 101-109.

Morgan, W.J., 2006. *Why sports morally matter.* New York, NY: Routledge.

Munro, M., 2008. Dressing for success at the Olympics; is it "doping on a hangar" or is it just a swimsuit? Either way, Canada is hoping that a little technology will lead to a gold strike in Beijing [online]. 27 July. Available at: <http://search.proquest.com/docview/434667680?Accountid=6180> [Accessed 24 January 2018].

Novak, M., 1994. *The joy of sports: End zones, bases, baskets, balls and consecration of the American spirit.* 2nd edition. New York: Basic Books.

O'Connor, L.M. and J.A. Vozenilek., 2011. Is It the athlete or the equipment? An analysis of the top swim performances from 1990 to 2010. *Journal of Strength and Conditioning Research,* 25(12), pp. 3239–3241.

Pargament, K. I., 2013. Searching for the sacred: Toward a non-reductionist theory of spirituality. In: K.I. Pargament, J.J. Exline, and J. Jones, eds. 2013. *APA handbooks in psychology, religion, and spirituality: Vol. 1 context, theory, and research.* Washington, DC: American Psychological Association. pp. 257-274.

Partridge, B., 2011. Fairness and performance-enhancing swimsuits at the 2009 swimming world championships: The "asterisk" championships. *Sport, Ethics and Philosophy,* 5(1), pp. 63–74.

Persson, I. and J. Savulescu., 2015. Summary of unfit for the future. *Journal of Medical Ethics,* 41: 338–39.

Porter, J., 2009. Implicit religion in popular culture: The religious dimensions of fan communities. *Implicit Religion,* 12(3), pp. 271-280.

Prebish, C. S., 1993. *Religion and sport: The meeting of sacred and profane.* Westport, CT: Greenwood Press.

Price, J.L., 2000. An American apotheosis: Sport as popular religion. In: B.D. Forbes and J. H. Mahan, eds. 2000. *Religion and popular culture in America.* Berkeley: University of California Press. pp. 195-212.

Price, J.L., 2006. *Rounding the bases: Baseball and religion in America.* Macon, GA: Mercer University Press.

Sanford, A.W., 2007. Pinned on karma rock: Whitewater kayaking as religious experience. *Journal of the American Academy of Religion,* 75(4), pp. 875–95.

Schultz, A.M. and P.E. Carron., 2013. Socratic meditation and emotional self-regulation: Human dignity in a technological age. *Journal of Interdisciplinary Studies,* 25(1-2), pp. 137-160.

Sherwin, S., 2007. Genetic enhancement, sports and relational autonomy. *Sport, Ethics and Philosophy,* 1(2), pp. 171–80.

Simon, R.L., 2001. Good competition and drug-enhanced performance. In: W. J. Morgan, K. V. Meier and A.J. Schneider, eds. 2001. *Ethics in Sport.* Champaign, IL: Human Kinetics. pp. 119-129.

Simon, R.L., 2004. *Fair play: The ethics of sport.* 2nd edition. Boulder, CO: Westview Press.

Smart, N., 1996. *Dimensions of the sacred: An anatomy of the world's beliefs.* Berkeley: University of California Press, 1996.

Tamburrini, C. M., 2007. What's wrong with genetic inequality? The impact of genetic technology on elite sports and society. *Sport, Ethics and Philosophy,* 1(2), pp. 229–38.

Tamburrini, C. M. and T. Tännsjö., 2011. Enhanced bodies. In: J. Savulescu, R. Tur Meulen, and G. Kahane, eds. 2011. *Enhancing Human Capacities.* Oxford, UK: Wiley-Blackwell. pp. 274-290.

Tännsjö, T., 2005. Genetic engineering and elitism in sport. In: C. M. Tamburrini and T. Tännsjö, eds. 2005. *Genetic technology and sport: Ethical questions.* London and New York: Routledge. pp. 55-69.

Trothen, T.J., 2015. *Winning the race? Religion, hope, and reshaping the sport enhancement debate.* Macon, GA: Mercer University Press.

Trothen, T.J., 2017. Moral bioenhancement through an intersectional theo-ethical lens: refocusing on divine image-bearing and interdependence. *Religions.* Special Issue: Religion and the New Technologies. Guest Editor: N. Herzfeld, [e-journal] 8(5), pp. 1-14. Http://dx.doi.org/ 10.3390/rel8050084.

World Anti-Doping Agency (WADA)., 2018. Prohibited list. Available at: <https://www.wada-ama.org/sites/default/files/prohibited_list_2018_en.pdf> [Accessed 23 January 2018].

Watson, N. J. and A. Parker., 2013. Sports and Christianity: Mapping the field. In: N. J. Watson and A. Parker, eds. 2013. *Sports and Christianity: Historical and contemporary perspectives.* New York: Routledge, 2013. pp. 9-88.

Zepp Staff., 2018. Zepp 2: Baseball support. Available at: <http://www.zepp.com/en-us/smartbat/> [Accessed 15 June 2018].

7.

Cyborg clergy and bionic Popes: An analysis of technological human enhancement from a Roman Catholic bioethical perspective

Michael Caligiuri

University of Manitoba

Abstract

Roman Catholicism holds a distinct place as an early foundation to the modern field of bioethics and maintains a long history of contribution when addressing ethical issues. As religious input to contemporary bioethical issues is required to ensure just representation and plural perspectives, a Roman Catholic approach to issues of human enhancement technologies allows for the study of a number of theoretical and present-day advancements which directly affect human form and function. Technological body modifications such as cybernetics and nanotechnology allow for an expansion of our inherent species-based norms and, as such, may present problems to equality, identity, privacy, and definitions of humanity itself. Ultimately, the validity of enhancement technologies must be addressed via religious interpretations in order to determine their moral worth and potential wide-scale acceptance. Within a Roman Catholic approach, this entails analyzing body modifications through a number of key sources such as: the natural law, Biblical input, contributions from Church Fathers, determinations of the magisterium, and modern concepts of social justice. By utilizing such sources, a Roman Catholic approach may recognize the diversity and current use of enhancement technologies and sanction their application within certain guidelines which protect key religious principles such as: prudence, stewardship, social justice, solidarity, human dignity, honesty, and a respect for future life.

Keywords: Bioethics, Cyborg, Modification, Enhancement, Technology, Clergy, Ethics.

Introduction

Analyzing a particular faith traditions' reaction to efforts to engineer human enhancement, this chapter surveys Roman Catholicism's response to enhancing cybernetic and nanotechnological body modifications. As Roman Catholicism has a long history of contributing to the field of bioethics and given the significant numbers of adherents worldwide, definitive policies on behalf of the Catholic Church in regards to technological human enhancement cannot be undervalued. Justifying this consideration, I first outline the value of religious input to bioethics, and present definitions of human enhancement and levels of normalcy in connection to cybernetic and nanotechnological implants to clarify ethical problems. This will include a proposed series of criteria to assess a technology's potential bioethical acceptability: implantability, permanency, power, and public interaction. This is followed by an analysis from a Roman Catholic perspective of select major social issues brought forward by enhancement technologies: commodification, eugenics, vulnerability, and distributive justice. Such issues and any medical-grade enhancements to the human body are judged by a combination of traditional and modern sources for Catholic bioethical ethical reflection. The present chapter concludes by assessing specific aspects of technological body modification which fall within acceptable limits of conventional Roman Catholic approaches to bioethics. The conclusion also addresses the areas of cybernetics and nanotechnology that would cause difficulty, if not prohibition. Through this analysis the basic questions of 'may I be a Catholic cyborg?' and (as they stem from laity) 'will there be a day when the clergy, from deacons to pontiffs, utilize cybernetic, bionic or enhancement technologies?' shall be clarified.

Why include a religious perspective?

There is a long history of religious opinion regarding the origins, composition and health of human beings that also includes official policy relating to areas of scientific research, medicine and technology. Generally, religious traditions address bioethical technologies in terms of moral acceptability with respect to doctrines and beliefs. Religious judgments are not restricted to medical interventions which threaten an individual's life, or specific health issues such as cases involving human reproduction; indeed, even cosmetic or artistic changes to the human body are often subject to religious guidelines (see Caligiuri 2013). Such guidelines allow us to categorize, if not predict, religious responses to cybernetic and nanotechnological body-modification processes. It is appropriate to include religious responses to medical and technological bioethical issues as the field itself derives from the combination of philosophical and religious sources or, as founder of the Kennedy Institute of Ethics and creator of the *Encyclopedia of Bioethics*, Warren Reich (2003, 185) labels them: the two legs of the humanities. Reich's endorsement of religious input to bioethical questions is shared by other scholars of the field. Indeed, the series of principles that serve as a standard for ethical deliberations in the life-sciences outlined in the 1978 *Belmont Report* (respect for autonomy, non-maleficence, beneficence, and justice), has arguably developed from religious foundations.[1] Yet the importance of a religious perspective to public debate of bioethical issues goes beyond historical applications or contributions to philosophical principles. Inclusion can be justified simply by virtue of liberty—to ignore the religious perspective dismisses billions of individuals who guide their life-choices by official religious policy. Catholic ethicist Lisa Sowell Cahill (2005) counters the marginalization of religious input in the public square by noting that policy-makers who exclude religious responses limit their own positions by assuming religious views bias bioethics (1).[2] Despite differing responses to specific bioethical dilemmas, there is consensus amongst many working within bioethics regarding the religious foundations of their field and the dangers associated with a purely non-theistic approach to all issues (see Curran 1982; Engelhardt, 2003). As political philosopher and Gifford Lecturer Jean Elshtain pointed out, religious contributions to bioethical problems have not died out, nor has an ethical framework based solely on scientism become the dominant form of social reflection (see Al-Hibri, Elshtain, and Haynes, 2001). The reality is that concerns over the limits of technology may be dubbed as hubris, or 'playing God', and are shared by people of all theological beliefs and none (President's Council on Bioethics, 323). As such, to be just during the analysis of a bioethical problem one must endeavor to include the perspectives of the great variety of beliefs that encompass the affected population.

Definitions of human enhancement and normalcy

When technology and enhancement merge on the domain of the human body encountering the term 'cyborg' is inevitable. Transitioning from a philosophical term related to guidance and control to the integration of the biological with the artificial (the mechanical and the computerized), scientists Manfred E. Clynes and Nathan S. Kline (1960) first used 'cyborg' in the September 1960 issue of *Astronautics* to describe a theoretical being, part human and part machine,

---

[1] Robert Veatch (2003, 69) notes that the ethical foundations of medicine first come from meta-ethical foundations, as does Jennifer K. Walter and Eran P. Klein (2003, ix).

[2] See also Bowman (2004), and Dumler (2003) on the critique of secular bioethics as presenting a predominantly white-male-oriented view to global issues, contributing to the marginalization of minority population by the use of utilitarian theories.

which could be developed for work in extra-terrestrial environments.[3] The exploration of the cyborg myth and image, including fractured identities, power and relationships in Donna Haraway's "A Cyborg Manifesto: Science, Technology, and Socialist-Feminism in the Late Twentieth Century" would consequently thrust the term into academic convention (see also Haraway, 1985). It is within Haraway's analysis that the distinctions between 'natural' (biological, not fabricated by humans) and 'artificial' (synthetic, manufactured by humans) blur, if not dissolve. This integration of the natural and artificial is thus a fundamental aspect of a cyborg as it provides the basis for an extension to human abilities, or enhancement.[4] Within the medical community, the term 'enhancement' is of less importance or esteem than 'treatment'. Enhancement is often equated with an extra, a bonus, a more-than-normal, or beyond-original level of human health and form. The primary goals of medicine are to save and prolong life, eliminate and prevent diseases, and reduce pain and sufferings; traditionally, enhancement encompasses that which goes beyond such goals, something that is less prioritized (particularly in resource-scarce, overburdened medical systems). This is not to say that, historically, certain types of enhancements have not been widely applied, primarily in the form of pharmaceuticals for altering mood, memory and attention spans or by similar brain modifying behaviors (Merkel et al., 147). However, unlike the cybernetic and nanotechnological options, these neuro/pharmacological enhancements do not entail permanent, physical or mental improvements in human performance or function so much as altered brain activity and abnormal sensations.

Director of the Center for Genetic Research Ethics and Law at the Hastings Centre, Eric T. Juengst (1998) gives a well-founded description of the uses for enhancement. He points out that the term is important within a bioethical discourse on two levels: the proper limits of biomedicine and the ethics of self-improvement (29). Juengst (1998) asserts that differing ethicists apply the term according to their respective fields of ethical analysis (44), and for biomedicine, enhancement can be differentiated from treatment in that enhancement encompasses "interventions designed to improve human form or functioning beyond what is necessary to sustain or restore good health" (29).[5] Hastings Centre fellow and ethicist Norman Daniels (2000) adds to this definition by

---

[3] As Clynes and Kline explain, "For the exogenously extended organizational complex functioning as an integrated homeostatic system unconsciously, we propose the term 'Cyborg'" (26-27).

[4] For more details on the development and application of the term see Caligiuri (2013).

[5] A definition used by the U.S. Presidential Council on Bioethics (2003).

including the limit of species-typical normal functioning or organization. This view of enhancement distinguishes it from (mere) treatment: that which restores but does not exceed species-based norms of human function, opportunity and range.[6] In this way, any intervention that expands upon one's functionality beyond a species-based norm would be classified as an enhancement.[7] Yet the application of species-based norms to the definition of enhancement does not necessarily fit all scenarios. For example, modern medicine accepts the beneficial use of vaccines and classifies them as valuable treatments, although they are not (in the strictest sense) part of our 'natural' makeup insofar as we are born without a great many. As such, vaccines are 'enhancements' for they provide an addition to the original state, added protection. So too may be corrective eye-wear, hearing aids, and steroid therapies, although such interventions are generally bound by what is considered to be species-based normal function. In the end, the true distinctions will depend upon how a technology is applied; what may be more contested is how to deem the end result as 'normal'.

Renowned sociologist Émile Durkheim (1982) notes that the most common behavior in a society is considered 'normal'; as a qualitative standard, normality can be defined by society as that which is considered average, ordinary, or within acceptable current conventions. Thus, 'normal' must be framed within an understanding of a given society, context and time. That which is considered normal today may not have been considered so in the past, nor assumed to be so in the future. Indeed, medical anthropologist Linda Hogle (2005) explains that the concept of 'normal' is a relatively recent addition to the English language and was adapted for biological use only around the mid-nineteenth century (697). Within biomedicine, 'normal' is a label and a quantitative measure, used to judge against preconceived or naturally occurring standards. As such, medicine defines what is normal within a measurable deviation and applies the data to a wide variety of physiological and psychological models. Baseline comparisons have provided standards of normal blood pressure, normal heart-rate, normal patterns of human development, normal height and weight, etc. All are averages, not necessarily universal, but extremely useful for controlled determinations of human health. As anthropologist and religious scholar Thomas Csordas (1998) notes, "medical science has been able to

[6] Interestingly, this definition may also hold that a treatment may be classified as an intervention that reduces, rather than adds to, a species-based norm of function.

[7] For a definition of enhancement that avoids the use of norms, see LeRoy Walters and Julie Palmer (1997) who explain that treatments apply to health-related issues, enhancements do not. See also Parens (1998).

conduct empirical study of the body, yielding statistical standards defining the 'normal' human body and methods by which medicine can manipulate and control bodies that diverge from those norms" (84). Hogle (1998) adds the further one is away from the hypothetical 'norm' the more one is to be classified as abnormal (698). This statistical model of normal is used by governments in managing a population's health and labor, and represents a political-juridical-institutional state of normalcy (Davis, 1995, 107). Thus 'normal' is a combination of government and medical guidelines as judged by society at a given time (regulations which vary across borders).[8] The ethical challenges associated with such a definition of normal include the validity of authorities to define standards and the marginalization of those deemed "abnormal." By classifying or "normalizing" the public and defining an individual's rights and privileges upon statistical models, we may risk more than meets the eye. Trends to classify segments of the populations as abnormal lead to classic ethical debates regarding eugenics, the rights of peoples with disabilities, and the value of human life. As ethicists and philosopher Anita Silvers (1998) explains, the distinction between treatment and enhancement inadvertently valorizes 'the normal' – or perhaps more precisely, valorizes those who are above 'the normal' and demotes those who do not live up to the statistical model—and in so doing, we inevitably segment society. One may argue that as enhancement technologies seek to add to species-based levels of normalcy they fall outside of such an argument, yet the social segmentation remains. Indeed, it is increased as there must now be a distinction between individuals who may be classified as: (1) normal, (2) abnormally low, and (3) abnormally high.

Assessing a technological modification

As I shall outline, assessing an enhancement technology can include religious guidelines which may entail great diversity given global plurality and differences in doctrines, yet there are certain criteria that may act universally. The quest for human enhancement is ultimately aimed at improving quality of life and is thus laudable, a characteristic shared with most of the life sciences. A technological enhancement becomes a cause for concern only when a number of specific criteria are met: implantability, permanency, power, and public interaction. A technology which fulfills all these criteria can be deemed a greater ethical risk as compared with a technology which does not. Implantability refers to direct attachment to or within the human body (such as a hearing aid or an artificial

[8] See for example Kohrman (2003) who notes that the Americans with Disabilities Act in the United States and the National Survey of Disabled Persons in China may thus define who is normal or disabled and funded care options.

heart). These devices include a great spectrum of modern medical designs such as valves, rings, grafts, meshes, stents, and patches. Certainly, implantability is associated with medical risk, the degree varying upon the substances, depth, and tissue types. Indeed, no implantable device is without risk for throughout their use, from initial manufacturing, insertion, and maintenance, there will always be associated chances of failure or rejection.[9] More so than implantability, dependent upon type and function, an enhancement technology may be permanent. Permanent devices require high levels of skill to implant, are not meant for removal and may need constant maintenance, hence entailing greater risk. An artificial organ such as a heart, pancreas or lung would be an example of a permanent device, since removal would require direct medical intervention and a high level of risk to the individual.[10] The criteria of permanency also entails the loss of a function if the device or implant were to be removed (possibly resulting in disability or death). Apart from the risks entailed by their removal, permanent body objects also introduce issues related to replacement, cost-burdens and the use of scarce resources.

The criterion of power is defined or measured on a scale relative to the norm of human function as outlined by Sabin and Daniels. The power of an enhancement technology or device may be determined by whether it provides capabilities or functions that are equal to or greater than species-based levels of normalcy and is best applied to devices or technologies on a case by case level. Truly, almost any tool could be labeled as "powerful" under the rubric of adding to human functionality (a hammer, for example), yet ethicists tend to bound this criterion within the context of implantability and permanency. For example, the inverted-series-connected (ISC) biomorph actuation devices discovered by Rutgers University-New Brunswick engineers: nano-scale stacks of molybdenum disulfide capable of actuation and lifting 150 times their own weight.[11] Such nano-materials may mimic biological muscle tissues, but would far exceed our species-based norms. While bioethical reflection upon any

---

[9] For specific examples of risk and medical device recalls, see the U.S. Food and Drug Administration Alerts and Notices wherein possible contaminations, malfunctions, and defects in currently used implantable devices are found.

[10] Devices which are permanently used in the human body are generally designed from materials which are bio-neutral so as to avoid infections or other risks to the user.

[11] Nanotubes allow for the sensations of touch and temperature for artificial skin, and because carbon nanotubes, 1/10,000 as thick as a human hair, are the most efficient thermal and electrical conductors known, they may be formed into water-resistant skin composites shaped by lasers (Fischman, 2010, 51). Nano-skin, combined with bionic prostheses, thus produces an artificial construct which is indistinguishable in terms of sensory information provided from their biological counterparts.

source of power is prudent, the greater the power of an enhancement technology or device, the greater the need for analysis of its impact upon equality, justice, and prudence. Finally, the criterion of public interaction measures an enhancement technology via interactions with other people and the environment. If an enhancement device has inherent characteristics which go beyond the confines of an end-user, either by way of its specific ability or by way of its use, then the category of public interaction is fulfilled. Implants or devices that monitor the public sphere, record, identify or otherwise influence the end-user in public beyond species-based or social norms meet this criterion. As such, issues of privacy, safety, and distributive justice arise when assessing the social implications of such devices. Additionally, it is important to note that although the public serves to shape the guidelines which may regulate such devices, the record of doing so with respect to technologies displaying the criteria of public interaction is rather poor. Novell technologies which integrate and monitor beyond the end-user are abundant, from the Google Glass optical head-mounted display developed by X (previously Google X) to Amazon's Echo and Alexa family of home assistants. The rapid consumption of these and similar products indicate a public that is less concerned with possible underlying issues of public interaction and more interested in the enhancements such technologies offer to daily life. Functioning as monitoring and identification systems, the ethical concern is that levels of integration reach beyond an end-user's understanding (or interest); linking meta-data across a wide-range of social-media platforms and location services, in essence following the end-user and nearby unaware individuals, and providing personal details to third-parties without counsel.[12] In the end, the more criteria that an enhancement device or technology fulfills, the more worthy it may be of regulation and, most certainly, ethical deliberation (cf. Caligiuri, 2013).

A religious perspective

Roman Catholicism has a rich history of ethical thinking and has articulated a well-defined set of ethical parameters that address biotechnological advances. Roman Catholicism's distinctive branch of moral theology was heavily influenced by classical Greek and medieval philosophers; indeed, *theologia moralis* has mostly been in use since the end of the sixteenth century.[13] Tracing

---

[12] For a detailed outline of the four criteria see Caligiuri (2013), Table 1.

[13] See Mahoney (1989). Early examples of specific religious considerations of health care topics, such as the duties and obligations of those acting as doctors and caregivers, were put forward by medieval theologians such as St. Antonius

the Catholic approach to modern bioethics, moral theologian Charles Curran (2003, 117) points out that during the maturity of the field from 1950 to 1970, it was the constant changing of technology which caused Roman Catholicism to react and elaborate on doctrine. Technological advances such as organ transplantation and reproductive devices became matters of legal interest and public policy, pushing Roman Catholic bioethics to new levels. Initially, Roman Catholic bioethical approaches responded to advancements on a case-by-case basis; tackling artificial insemination, the contraceptive pill, or new forms of sterilization. Catholic ethicists such as Kevin Quinn and Richard McCormick note that this 'piece-meal' approach has given way to a new methodology within Roman Catholic bioethics, one which relies much more on the "the tenets of Catholic Social Teaching: common good, human dignity, option for the poor, and stewardship" (Daniels-Sykes, 2007, 7). As Sr. Shawnee Marie Daniels-Sykes (2007, 102) explains, the shift away from manuals and magisterium to modern approaches allows Roman Catholic bioethics a broader understanding of issues in relationship patient rights and patient autonomy. Yet this is not to say that the Roman Catholic Church has abandoned the input of sources outside of Catholic Social Teaching. As Catholicism draws on 2000 years of theology and moral philosophy it has acquired several important tools which act in conjunction when addressing bioethical issues. The relevant sources include, but are not limited to: 1) natural law; 2) Biblical guidelines; 3) contributions from Church Fathers; 4) input from the magisterium, and; 5) modern concepts of social justice. In applying these foundations to modern bioethical issues, Catholicism displays a characteristic fusion of continuity and change, since many of the answers to ethical dilemmas draw on both historical precedents and contemporary Church pronouncements.

The concept of natural law has roots as far back as Plato, the Stoics, the first Christian Apostles and ancient philosophers, yet it continues to be used in the present by theologians and legal analysts. Natural laws are the innate rules of the world which are differentiated from human-made laws and divine law. Hearkening back to Paul's first-century Epistle to the Romans, natural law is understood as accessible to all people:

> When Gentiles, who do not possess the law, do instinctively what the law requires, these, though not having the law, are a law to themselves. They show that what the law requires is written on their hearts, to which their own conscience also bears witness; and their conflicting thoughts will accuse or perhaps excuse them. (Romans 2: 14-15)

(Antoninus), the fifteenth-century Archbishop of Florence, leading to the inclusion of regulations regarding competence, diligence and honesty in Canon Law.

Paul is speaking here of the pagans, who although they do not have the traditional Mosaic law, do have another law which teaches them the difference between right and wrong—a law which is fundamental to all persons, ingrained by our natures and gives people a knowledge of good and evil. Catholicism views natural law as universal, extending to all persons regardless of religious affiliation and it teaches that positive laws, or human laws, are meant to model themselves upon a set of principles that are the highest expression of nature and reason, which is God. How one is to perceive the natural law and discover its guidance in relation to the positive law is something that is often ascribed to human intuition.

At the heart of Christianity are the general principles espoused in the Bible which form the rudiments of Catholic bioethical policies: sanctity for life, love of all persons, and stewardship of the body. While many specific verses found in the Scriptures hold to such principles, no single passage may be adequately applied to all bioethical dilemmas. Roman Catholic approaches to bioethics tend to apply Biblical contributions generally rather than specifically. This is, in part, due to Catholicism's recognition of the importance of the text's historical context and intention, differentiating between narrative or prosaic language and literal interpretation. This reliance on both established and broad interpretations of Biblical texts is reflected in many Roman Catholic statements, particularly the Second Vatican Council's, *Dei Verbum* (1965).

While Christian theologians through the centuries commented on a variety of ethical issues, many contemporary bioethical dilemmas borne from new technologies are obviously not directly addressed in their works. This is not to say that such biotechnological questions could not be analyzed via broad interpretation or using parallel analogies. Rather, it is a recognition that modern technological issues simply lack specific reference in patristic and medieval sources. However, this lack of specificity does not necessarily hinder their importance; for many of the foundations to moral theology are still applicable and are used in Catholic bioethical responses. One of the most prominent theologians to contribute to the foundations of bioethics is Thomas Aquinas. Through his extensive writings in the *Summa Theologica* and the *Summa Contra Gentiles*, a number of philosophical and theological positions regarding human beings are thoroughly discussed. Questions ranging from ownership and stewardship to prescribed guidelines on the body's constitution and care can all be found in subsections of his *Summa Theologica*. Aquinas's principles to Catholic ethics became determinants in questions on organ transplantations, amputations, and even capital punishment. Such principles include: The Principle of Totality—an opinion regarding the relationship between individual body parts and the whole; Stewardship—a concept which holds order to all things in the universe and, to this end, the limitations which a person has in decisions over their own body; the Principles of Integrity and Mutilation—the limitations and morality of separating principle parts of the human body from others.

Contributions on the part of the magisterium come in a variety of forms: apostolic letters, encyclicals, addresses, and even interviews for various forms of media. Popes and councils present teaching on broad bioethical issues and, occasionally, specific cases which display moral dilemmas. While no specific statement on bioethics has been classified as infallible under the guidelines set out by the First Vatican Council they are, nonetheless, considered significant contributions which help shape Church policy. As the magisterium has responded to differing bioethical issues through the years with reference to their specific historical and social circumstances, a complete list of individual statements is beyond our scope. Yet, to cite one example of papal input, between 1942–1958 Pius XII contributed more than 80 individual letters and addresses to differing medical societies and specialties with regards to Catholic morality and human health, plus the encyclical (1943) *Mystici Corporis* ("On the Mystical Body of Christ"). Further, several popes founded the Pontifical Academies to provide autonomous research and continuous study on academic topics of importance. These academies are dedicated to specific fields: Science, Social Sciences, and Life, each instituted (or rededicated) by popes interested in furthering the relationship between Catholicism and academic disciplines. For example, the Pontifical Academy of Life (PAL) (established 1994) publishes a yearly Plenary Assembly declaration as well as reflections on papal encyclicals, joint-statements and summaries on various bioethical topics.

As noted, today Catholic social justice teaching underpins many responses to modern bioethical issues. Catholic appeals to the concept of social justice have a long-standing history (perhaps best articulated by Pope Leo XIII's (1891) encyclical *Rerum Novarum),* and always include recourse for a preferential option for the poor, distributive justice, human rights, personal dignity and ecological stewardship. Alongside other important teachings, these ideas are outlined in magisterial documents such as *Gaudium et Spes* (Second Vatican Council, 1965), *Pacem in Terris* (John XXIII, 1963), and *Economic Justice for All* (USCCB, 1986). For the Roman Catholic Church, a bioethical issue must be examined on social levels in order to ensure a spirit of safety, uphold equality, and benefit the common good. Without such input, a scientific advance may endanger individuals and the public, creating disparity on economic and social levels which oppress or even kill. Catholicism utilizes several principles of social justice when considering ethical issues and is able to organize on differing levels to carry out action. On a basic level the faithful are called upon to live in solidarity and with due consideration to the social needs of others; thus, Catholics should avoid a bioethical technology which the magisterium has deemed immoral and reject options contrary to Catholic teachings. In accord with this approach, the Roman Catholic Church endorses an educational aspect of social justice in order to inform members and non-members about social issues facing the world. Moreover, on a level of community organization, it actively encourages peoples to gather together so as to discuss and advocate for just action on issues which may have negative consequences to society. This power to organize can be an effective tool in changing government policy and in setting community standards.

As technological enhancements offer the ability to exceed species-based levels of normal human function, a number of social issues immediately become linked. Chief among these concerns, and particularly relevant to Roman Catholic bioethical approaches, are: the reduction of the human body to constituent commodities; social concerns arising from the disparity of function between enhanced and non-enhanced individuals, a devaluing of persons without full or enhanced functionality; and ethical issues arising from the quest for human perfection through the body. As enhancement technologies directly involve the human body and require advanced, highly sophisticated equipment, a looming moral problem is that of commodification—the reduction of life to measurable economics. The problem has been a focus of Catholic Church teaching for several decades, born from the debates on reproductive technologies and the marketplace for human organs and addressed in a 2003 Pontifical Academy for Life communiqué (cf. John Paul II, 1991). The expansion and innovations of technologies have caused bioethicists to focus on commodification and the increase of cultures "preoccupied with commercial consumption" (Wright, 2000, 10). Arguing that human culture has already embraced a reductionist view of life, Indiana University professor of law R. George Wright (2000) explains the link between commodification and human dignity: "the commodification of life, along with overly reductionist views of life and culture, do not bode well for the deep respect of human dignity in the future. There is ultimately no reason to accord genuine respect to mere commodities, or to mere mechanical objects, however complex or high their market price. Commodities generally do not possess dignity in the sense classically ascribed to humans or to rational persons. Nor is it at all clear why humans would really possess dignity in this sense if being human were thus reducible. We can certainly admire the sophistication of a piece of computer hardware or of a software program. But we do not ascribe genuine dignity to either, or even to their combination" (12). Examples of the type of reductionism that Wright warns of are found through 'transplant tourism' wherein the wealthy seek live black-market human organs from the poor in places such as India, providing a few thousand rupees for the 'donor' kidney, liver or lungs while 'brokers' keep thousands more in fees. A Catholic rejection of reductionism upholds the body as both good and admirable, while at the same time insisting that individuals are "more than a speck of nature," more than the "sum of mere things" (Paul VI, 1965, #14). It is also a feature of the progression in Catholic Social Teaching from Leo XIII's *Rerum Novarum* (1891) to John Paul II's *Centesimus Annus* (1991) wherein concerns have expanded beyond charity to include justice. This refocusing on the poorest segments of society is a theme of Roman Catholic Social Teaching and results in apprehension towards advancements which could lead to further disparity between all peoples and disadvantages to the underprivileged.[14]

As enhancement technologies offer an array of benefits beyond species-based norms, they inherently hold great value and bring forward economic issues on social and individual levels. Socially, body enhancements act as a new commodity for markets with a corresponding influence on the businesses of medicine, research and development, insurance, manufacturing, advertising, and law. Individually, body enhancements entail material goods and the required services for their installation and maintenance. Current market trends indicate that the sale of body parts such as organs is a booming global business; indeed, the World Health Organization (2017) estimates that 10% of transplants worldwide are organs that have been sold to the recipient. As Hogle (2015) and others have noted, engineered artificial tissues capable of enhancement constitute a different sort of economic problem, where supply is theoretically readily available yet possibly restricted by politics and market regulations on the trade of human materials (see Franklin, 2001; Waldby, 2002). As Hotze et al. (2000), conclude in their enhancement technologies survey utilizing a random sample of 1500 physicians, if such interventions were to become available by prescription tomorrow, inequalities in access to them would arise both on the basis of insurance coverage and ability to pay and also because of variation in physician willingness to prescribe them. Both of these factors would likely exacerbate existing social disparities across racial, ethnic, economic, and other socio-demographic lines, as already advantaged groups are better able to actively seek physicians willing to prescribe them enhancements and to pay for these interventions out of pocket (11).

---

[14] As expressed in *Gaudium et Spes*, "A consistent theme of Catholic Social Teaching is the option or love of preference for the poor. Today, this preference has to be expressed in worldwide dimensions, embracing the immense numbers of the hungry, the needy, the homeless, those without medical care, and those without hope" (Paul VI, 1965, #42).

Catholicism addresses the body and economics on both individual and social levels and has maintained a consistent policy of declaring the body and all its parts as priceless. This is a rejection of the devolution of the person to an object, a collection of parts, rather than an incorporated subject. On social levels, the Roman Catholic Church maintains health is a universal good to be defended, not commodified, and organizations offering health services must "rethink their particular role in order to avoid having health become a simple 'commodity', subordinate to the laws of the market, and, therefore, a good reserved to a few, rather than a universal good to be guaranteed and defended" (Benedict XVI quoted in CNS Staff, 2012). Additionally, papal pronouncements charge that consumerism is a path towards dissatisfaction. As John Paul II (1987) explained: "Excessive access to all kinds of things—sometimes called consumerism—enslaves people and does not make them happy. The more one possesses, the more one wants, while the deeper human hopes remain unsatisfied and even stifled" (#28). This is a continuation of the themes of Vatican Council II, which stressed that people are worth more than their possessions and, ultimately, that "[t]echnical progress is of less value than advances towards greater justice, wider kinship and a more humane social environment" (Paul VI, 1965, #35). In order to remain consistent with previous judgments and maintain a focus on the preferential option for the poor, a Catholic approach to enhancement technologies will reject a market that prioritizes upgrades, interchangeable implants, and an industry based upon an economy of increased body function.

As eugenics aims to reshape the human condition by the elimination of physical and mental shortfalls, it connects to a number of technological enhancement issues as well as topics of social justice. For example, eugenics links back to our definition of 'normal', in that governments have authority over medical treatments based upon the system of classifying human deviations from bodily norms. Eugenics is further linked with vulnerability and, in turn, with human dignity and functionality; for when our capabilities are refereed by biological standards, segments of the population will inevitably be labeled as inferior to the norm. Thus, eugenic practices hold commonality with technological body enhancements and, despite the theoretical goal to help eliminate human suffering, carry a long history of bioethical menace.

The resurgence of eugenics via genetic manipulation has not escaped the notice of either bioethicists or theologians. Examples of Roman Catholic bioethical sources which have responded to historical and new eugenic movements include papal encyclicals and magisterial proclamations such as *Casti Connubii* (Pius XI, 1930), *Evangelium vitae* (1995, #14 & 63), *Donum vitae* (1987) and *Dignitas personae* (2008), as well as denunciations of the technology at the PAL general assemblies—as in 1998 when it met to discuss the implications of the Human Genome Project. While early papal denunciations on the part of Pius XI and Pius XII addressed the theoretical application of eugenics borne out of Darwinian sociology, a modern Catholic approach focuses on all processes which attempt to perfect human nature through biomedicine. As the process of manipulating genes to achieve a "new eugenics" imperative has been conceded by the magisterium as inherently risky and a violation of the principle of stewardship, it is realistic to assume that cybernetic or nanotechnological processes would be the same. Roman Catholic bioethics answers that a push towards perfection, even with the utmost sincerity and possibility of achieving a better society, does not genuinely consider the purpose of a human being—that accomplishment of "moral perfection [proportional] to his potentialities and his efforts, always assisted by the grace of God" (Chyrowicz, 2009, 15). Within Roman Catholic bioethical approaches, the nature of the body cannot be solely viewed as oriented towards a goal of earthly perfection.[15] Throughout history Catholicism has expressed a belief that defects of the body (be they physical or mental) are to be seen as signs of spiritual testing, prowess, and even gifts; to be authentically human is to be subject to vulnerability.[16] This is not to say that Catholicism endorses or somehow upholds suffering as a universal state, requirement or benefit. Rather, it recognizes that the quest to add function or abilities beyond the species-based norms diminish those who do not meet these norms to begin with. As noted, this approach to vulnerability stems from a number of sources. Certainly, scripture is a source of numerous examples of healing to overcome the vulnerabilities of body and spirit, while at the same time acknowledging finality to the body. Biblical examples of healing and human frailty are reflected in various Catholic directives for healthcare and medical-ethics programs, which take up themes found in Acts 3:16 and 9:34, Phil 2:26, 2 Tim 4:20, 2 Cor 12:10, James 5:115, etc. The sick are healed, through Christ, love, faith and the actions of his followers, who themselves are subject to infirmity or illness – a precarious and universal

---

[15] As Discalced Carmelite and philosopher Edith Stein (1987) observed, "a man's one-sided endeavor to achieve perfection easily becomes a decadent aspiration in itself; our desire for knowledge does not respect limits placed on it but rather seeks by force to go beyond these limits; human understanding may even fail to grasp that which is not essentially hidden from it because it refuses to submit itself to the law of things; rather, it seeks to master them in arbitrary fashion or permits the clarity of its spiritual vision to be clouded by desires and lusts" (70).

[16] See Teilhard de Chardin (1950); John Paul II's (1984) *Salvifici Doloris*; Kreeft (1986); Vanier (1998); Kane (2002); Garcia (2006).

characteristic of all people. And yet weakness can be embraced, suffering may necessitate patience, and one's distress is said to be shared by God. As the USCCB (2009) states in *Ethical and Religious Directives for Catholic Health Care Services*, "the mystery of Christ casts light on every facet of Catholic health care: to see Christian love as the animating principle of health care; to see healing and compassion as a continuation of Christ's mission; to see suffering as a participation in the redemptive power of Christ's passion, death, and resurrection; and to see death, transformed by the resurrection, as an opportunity for a final act of communion with Christ" (6). Catholic experts in the field and publications from the PAL extoll the virtues of vulnerability and emphasize the accessibility of moral good to all persons, the healthy as well as the disabled.[17] Indeed, many contemporary moral philosophers now believe that the connection between human vulnerability and concepts such as autonomy, justice, and social duty is so fundamental that it has become a cornerstone to bioethical problems and meta-ethical theory.[18] This is supported by data from surveyed physicians indicating that approximately one-third agreed with the statement that "[u]sing medicine for enhancement is wrong because some human suffering has value," even though a majority of respondents believed that "[t]he ultimate goal of medicine is to eliminate human suffering" and "...medicine for enhancement reduces human suffering" (Hotze et al., 6).

Traditionally, the Roman Catholic Church has advocated for the rights of the physically and mentally disabled as part of its mandate to protect all life despite age, form, or qualified levels of biological normalcy. By embracing human vulnerability as part of the complete human experience, the Roman Catholic tradition runs counter to philosophies which isolate and quantify life on levels

---

[17] Such a view of the value of vulnerability is not confined to a Roman Catholic ethos or even a religious ethos; as Reich points out many, many contemporary philosophers increasingly regard vulnerability as part of the basic identity of all humans (Reich, 1978, 184). Indeed, a list of those who incorporate human vulnerabilities on purely philosophical grounds includes Emmanuel Lévinas, Jürgen Habermas, Paul Ricoeur, Alasdair MacIntyre, Susan Okin, Robert Goodin, Leon Kass, and Judith Shklar. In their respective ways, each finds that human weaknesses and fragilities play a key role in what it is to be human and the subsequent responsibilities and virtues that must be taken on in order to best operate in a just society.

[18] Reich explains that the principle of vulnerability now belongs as an essential principle in bioethics (and bio-law) and he echoes MacIntyre's belief that "due to the thinness of the abstract principle of nonmaleficence and beneficence in contemporary analytic bioethics, these principles should be replaced with a more richly responsive virtue of concerned care" (Reich, 1978, 185).

of functionality, minimizing the relational value of persons regardless of ability. From a Catholic perspective, such a reductionism diminishes human dignity; it segregates and marks vulnerable persons as lower in value. Several Catholic moral philosophers speak to the fundamental importance of human mortality and warn that health care should not be focused on endeavors which seek unrealistic concepts of human existence. Indeed, the Roman Catholic Church has included such instruction as part of its Catechism. Yet, if morality requires respect for the life of the body, it does not make it an absolute value. It rejects a neo-pagan notion that tends to promote the cult of the body, to sacrifice everything for its sake, to idolize physical perfection or success at sports. By its selective preference of the strong over the weak, such a conception can lead to the perversion of human relationships.

Along with the aforementioned problems with human vulnerability and commodification, Catholicism looks with anxiety at enhancing devices that will cause social fragmentation, economic disparity, and possibly tyrannies (see Hook 2002; 2004). From a Roman Catholic bioethical approach, such problems fall within the scope of social justice on many levels: increased social division by way of unenhanced versus enhanced individuals; distributive and economic inequalities in medical care and a corruption of the role of physicians; and risks to individual personalities, mental processes and consciousness by way of addiction and manipulation. As a result of its focus on social teaching, an overarching theme of social justice permeates Roman Catholic bioethics, emphasizing the importance of living justly in community. A corollary to the Roman Catholic emphasis on human frailty and the limits of human life is the need for social caring and moral responsibility towards others. As Jesuit theologian Bernard Lonergan explains in *Insight*, freedom is always exercised in a matrix of human relationship, in community, because human beings have a primordial sympathy for one another—we do not live with one another as in an anthill but in relationship with feelings and commitments (Creamer, 1996, 84). Indeed, Lonergan explains that human progress is essentially and prominently a healing/creating process—on personal and social levels. To this end, the social teaching of the Catholic Church insists that the human community, including its governments, must be actively concerned in upholding the health and welfare of all persons, so that they may contribute to the common good of all. Such a teaching is encapsulated in the principle of the common good and requires respect for persons, social welfare, and amity. In his encyclical *Pacem in Teris*, Pope John XXIII (1963) defined the common good as "the sum total of social conditions which allow people, either as groups or as individuals, to reach their fulfillment more fully and more easily" (#55). Increasingly emphasized after Vatican Council II, Catholicism calls upon people to go beyond selfish interests and attend to others in a spirit of cooperation and love; or as explained in *Gaudium et Spes*, "...when the order of values is jumbled and bad is mixed with the good, individuals and groups pay heed solely to their own interests, and not to those of others. Thus it happens that the world ceases to be a place of true brotherhood" (Paul VI, 1965, #37). Roman Catholic ethicist and physician, Christopher Hook, warns that each of the dimensions which make up the common good are in danger from cybernetic enhancement technologies: the physical by way of the inherent risks to human health; the intellectual by way of their potential for corruption, their addictive-like dependence, and their unforeseen impacts upon our rational abilities and personalities; and lastly, the moral and ethical dimension by way of technology's propensity to isolate individuals, ruining relationships and offering an array of temptations (2004, 535). In defining the social problems of technological enhancements, Hook fears that some members of society will become "incrementally enhanced and plugged into cybernetic communities, these individuals will share less and less in common with the unenhanced, fragmenting society; potentially generating decreasingly compatible, or even competing, separate societies" (2004, 535).[19]

Hook (2004) raises a further social concern about cybernetic and nanotechnological enhancements by contending that they endanger the practice of medicine. Apart from unscrupulous doctors peddling enhancement technology, Hook fears that a fundamental change in the profession may occur, "from a group committed to healing (with a dominant ethos of beneficence in trust and nonmaleficence) to individuals skilled in surgical technique who are merely technicians" (Hook, 2004, 535). But the problem is not limited to a shift away from compassionate medicine to mere mechanics; it also causes an increase in demands from medicine. Hook views these demands as immoral in two respects: by encouraging unjust resource allocation and unaffordability, and by corrupting the healer-patient relationship. First, linked to the pressures of distributive justice and commodification, it is feared that medical resources may be squandered on frivolous requests for upgrades, implants, and technological abilities, instead of reaching the truly needy, the poor, and the underdeveloped.[20] Roman Catholic directives for health care services mandate working to ensure health care delivery systems provides adequate health care for the poor, with particular attention given to the health care needs of the uninsured and the underinsured (e.g., USCCB, 2009, 10). If enhancement technologies cause further limitations to the delivery of health services,

---

[19] Similar concerns have been given by Maguire and McGee (1999), Parens (1998), and Merkel et al. (2007), who fear that while technologies like neural implants can help restore function to the disabled, the same implants "could potentially lead to the formation of elites and considerable social bias" (Merkel et al., 155).

[20] See Hook (2004); Garcia & Sandle (2008); Mitchell et al., (2007).

distributive justice is violated. Catholicism's stance on such economic inequalities incorporates its longstanding tradition of the preferential option for the poor—that "special form of primacy in the exercise of Christian charity, [affecting] the life of each Christian inasmuch as he or she seeks to imitate the life of Christ, [and applying] equally to our social responsibilities and hence to our manner of living, and to the logical decisions to be made concerning the ownership and use of goods" (John Paul II, 1987, #6).

Roman Catholic guidelines for enhancement technologies

The use of technology as an enhancing tool does not in itself constitute a bioethical or moral concern for the Roman Catholic tradition; nor is the aspiration for human self-improvement or creative self-transformation. Assuming a minimal risk to health and an option to reverse the applications of enhancement if needed, it is unlikely a Roman Catholic bioethical approach would fault cybernetic and nanotechnological implants. However, a problem remains in their potential to interrupt just levels of social equality. As well, while physical and mental function can increase with enhancement, there are no easy methods to enhance those human characteristics which Roman Catholicism stresses as more valuable to persons: the virtues. To date, no enhancement device has been developed to increase one's level of compassion, one's solidarity with others, or one's moral compass. As such, according to a Roman Catholic bioethical approach, cybernetic and nanotechnological enhancement technologies appear aimed at the lesser qualities of people, qualities which the tradition does not necessarily even define as requirements for life and its protection. In essence this may be classified as a lack of focus, in so far as the resources dedicated to developing enhancement devices could be aimed at equalizing the current disparities between individuals rather than by creating new ones. This holds with the importance that the Roman Catholic Church places on authentic human development, as social "development cannot occur unless individual men and their associations cultivate in themselves the moral and social virtues, and promote them in society; thus, with the needed help of divine grace men who are truly new and artisans of a new humanity can be forthcoming" (Paul VI, 1965, #30).

The concerns are deep, and from a Roman Catholic bioethical perspective there are dangerous waters within the ocean of potential enhancement technologies. Yet, this does not mean that a Catholic response would forbid cybernetic or nanotechnologies aimed at human enhancement. It simply means that a moderate approach may be the way forward. For example (as given by Caligiuri, 2013), the following set of guidelines would respect the continuity of Catholic positions regarding previous technological advancements and addresses most (if not all) of the established outlined ethical concerns:

1. In keeping with the principle of stewardship and virtues such as prudence, an enhancement technology must not cause excessive risk to human health or the ability to return individual body parts to a state of natural function. This includes risks in altering the existing consciousness of the individual and electronic methods of control by way of neural enhancement.
2. In keeping with social justice, the economic burden of an enhancement technology must be weighed against the greater needs for one's self and the community.
3. In keeping with a spirit of human dignity, an enhancement technology must never reduce the body to levels of commercialization or commodification. The inherent worth of a body's function (even enhanced function) cannot be equated to the value of personhood.
4. In keeping with principles of solidarity and the common good, an enhancement technology must not detach an individual from the greater community, nor should it allow for a violation of other's privacy.
5. In keeping with the principle of the sacred gift of life, an enhancement technology must not interfere with the natural course of human reproduction, nor be applied in any form to alter the natural genetic makeup of a person, nor attempt to thwart a natural end to life.
6. In keeping with a spirit of authenticity and the virtue of honesty, an individual's intent to enhance must not stem from immoral motives, deceit, or frivolity.

Given these collected guidelines, many of the concerns regarding health and social issues with cybernetic and nanotechnological body modifications may be mollified, if not abated. In its reflection on an ethical vision for biomedical research, the Pontifical Academy for Life (2004) noted that "...in principle... there are no ethical limits to the knowledge of the truth, that is, there are no 'barriers' beyond which the human person is forbidden to apply his cognitive energy: the Holy Father has wisely defined the human being as '*the one who seeks the truth*' (*Fides et ratio,* n. 28); but, on the other hand, precise ethical limits are set out for the manner the human being in search of the truth should act, since '*what is technically possible is not for that very reason morally admissible*' (Congregation for the Doctrine of the Faith, *Donum Vitae,* n. 4)." From a Roman Catholic perspective, the application of enhancement technologies is not inadmissible unto itself, yet, methods matter. Without ethical limitations to guide their implementation, society risks devaluing segments of itself and reducing the person to a set of controlled devices which enhance function but not substance, form but not soul. The reality is, given the widespread use of artificial organs and engineered restorative devices such as cochlear implants, artificial retinas, and (the now common) pacemakers, there is no question that types of cyborgs walk amongst us today. As noted, this is simply due to the very real integration of the synthetic with the biological; or as Haraway (1995) explains, "Modern medicine is...full of cyborgs, of couplings between organism and machine, each conceived as coded devices..." (149-150). Thus the question, "May I be a Catholic cyborg?" may already be answered in the affirmative—there exist millions of technological end-users joined in a variety of ways with devices that enhance their quality of life. As such, there are indeed cyborg clergy; serving their faith-communities around the world, which are also incorporative of cybernetic, bionic, and enhanced beings. At this point, the only true distinction may be, will the integration and function of our cyborg selves seek to implement higher states of enhancement that go beyond our species-based norms? Given the concerns of a Roman Catholic approach to bioethics, such a pursuit must address the needs of others: economic, social and ethical risks, coupled with an understanding of our personal spiritual development. If these needs cannot be reconciled with a desire to enhance beyond the norm, it is unlikely that cybernetic or nanotechnologies would truly add to an authentic advancement of humanity.

References

Acerce, M., E. Koray Akdoğan, and M. Chhowalla., 2017. Metallic molybdenum disulfide nanosheet-based electrochemical actuators. *Nature*, 549, pp. 370-373.

Al-Hibri, A.Y., J.B. Elshtain, and C.C. Haynes., 2001. *Religion in American public life: Living with our deepest differences.* New York, NY: W.W. Norton and Company Inc.

Bowman, K., 2004. What are the limits of bioethics in a culturally pluralistic society? *Journal of Law, Medicine & Ethics*, 32(4), pp. 664-669.

Cahill, L.S., 2005. *Theological bioethics: Participation, justice, change.* Washington, DC: Georgetown University Press.

Caligiuri, M.J., 2013. *Traditional and new enhancing human cybernetic and nanotechnological body modification technologies: A comparative study of Roman Catholic and transhumanist ethical approaches.* Ph.D. University of Ottawa.

Clynes, M. and N.S. Kline., 1960. Cyborgs and space. *Astronautics*, 26(27), pp.74-75.

Chyrowicz, B., 2009. The risks of pushing towards perfection. Proceedings of the pontifical academy for life on the occasion of its 15th general assembly: "New frontiers of genetics and the dangers of eugenics." Rome: *L'Osservatore Romano.*

CNS Staff., 2012. Health is universal good to be defended, not commoditized, says Pope. *Catholic News Service* [online]. Available at:

<http://www.cathnewsusa.com/2012/11/health-is-universal-good-to-be-defended-not-commoditized-says-pope/> [Accessed 30 November 2012].

Congregation for the Doctrine of the Faith., 1987. Instruction on respect for human life in its origin and on the dignity of procreation — *Donum Vitae*. The Holy See.

Congregation for the Doctrine of the Faith., 2008. *Dignitas Personae*: On certain bioethical questions. The Holy See.

Creamer, D., 1996. *Guides for the journey: John Macmurray, Bernard Lonergan, and James Fowler*. Lanham, MD: University Press of America.

Csordas, T.J., 1998. Body. In: W.T. Reich, ed., *The ethics of sex and genetics: Selections from the 5 volume Macmillian Encyclopedia of Bioethics*. Revised edition. New York, NY: Macmillan Reference. pp. 84-91.

Curran, C.E., 1982. *American Catholic social ethics: Twentieth-century approaches*. Notre Dame, IN: University of Notre Dame.

Curran, C.E., 2003. The Catholic moral tradition in bioethics. In: J.K. Walter and E.P. Klein, eds. *The story of bioethics: From seminal works to contemporary explorations* Washington, DC: Georgetown University Press, pp. 113-130.

Daniels, N., 2000. Normal functioning and the treatment-enhancement distinction. *Cambridge Quarterly Healthcare Ethics*, 9(3), pp. 309-322.

Daniels-Sykes, S.M., 2007. *Roman Catholic social bioethics critiques secular bioethics: Fetal tissue research and vulnerable populations*. Ph.D. Marquette University.

Davis, L., 1995. *Enforcing normality: Disability, deafness and the body*. New York, NY: Verso.

Dumler, N.L., 2003. *The harm of neglecting embodiment: How biomedical ethics' neglect of bodies and context hurts women and minorities*. M.A. University of Tennessee, Knoxville.

Durkheim, E., 1982. *Rules of sociological method*. Translated by Macmillian Press Ltd. New York, NY: The Free Press.

Engelhard, T.H., 2003. The foundations of bioethics: Rethinking the meaning of morality. In: J. Walte and E. Klein, eds. *The story of bioethics: From seminal works to contemporary explanations*. Washington, DC: Georgetown University Press. pp. 91-109.

Fischman, J., 2010. The blind can see: A one-armed woman can fold her shirts. *National Geographic*, 217(1), pp. 34-54.

Franklin, S., 2001. Culturing biology: Cell lines for the second millennium. *Health*, 5(3), pp.335-354.

Garcia, J.L., 2006. Sin and suffering in a Catholic understanding of medical ethics. *Christian Bioethics*, 12(2), pp. 165-186.

Garcia, T. and R. Sandler., 2008. Enhancing Justice? *Nanoethics*. Department of Philosophy and Religion Nanotechnology and Society Research Group. Northeastern University.

Haraway. D., 1985. Manifesto for cyborgs: Science, technology and socialist feminism in the 1980s. *Social Revue*, 80, pp. 65-108.

Hogle, L, 2005. Enhancement technologies and the body. *Annual Review of Anthropology*, 34, pp. 695-716.

Hook, C.C., 2002. Cybernetics and nanotechnology. In: J.F. Kilner, C.C. Hook, and D. Uustal., eds. *Cutting-edge bioethics: A Christian exploration of technologies and trends.* Grand Rapids, MI: W.B. Eerdmans Publishing, pp.52-70.

Hook, C.C., 2004. Techno sapiens: Nanotechnology, cybernetics, transhumanism and the remaking of humankind. In: C.W. Colson and N.M.S. Cameron, eds. *Human dignity in the biotech century: A Christian vision for public policy.* Downers Grove, IL: intervarsity Press, pp. 75-97.

Hotze, T.D., K. Shah, E.E. Anderson, and M.K. Wynia., 2011. Doctor, would you prescribe a pill to help me…? A national survey of physicians on using medicine for human enhancement. *The American Journal of Bioethics,* 11(1), pp. 3-13.

John XXIII., 1963. Encyclical letter *Pacem in Terris* on establishing universal peace in truth, justice, charity, and liberty. Available at: <http://w2.vatican.va/content/john-xxiii/en/encyclicals/documents/hf_j-xxiii_enc_11041963_pacem.html> [Accessed 6 July 2018].

John Paul II., 1987. Encyclical letter *Sollicitudo Rei* Socialis—The social concern. Available at: <http://w2.vatican.va/content/john-paul-ii/en/encyclicals/documents/hf_jp-ii_enc_30121987_sollicitudo-rei-socialis.html> [Accessed 6 July 2018].

John Paul II., 1991. Encyclical letter on the hundredth anniversary of *Rerum Novarum—Centesimus Annus.* Available at: <http://w2.vatican.va/content/john-paul-ii/en/encyclicals/documents/hf_jp-ii_enc_01051991_centesimus-annus.html> [Accessed 6 July 2018].

John Paul II.,1995. Encyclical letter the gospel of life—*Evangelium vitae.* Available at: <http://w2.vatican.va/content/john-paul-ii/en/encyclicals/documents/hf_jp-ii_enc_25031995_evangelium-vitae.html> [Accessed 6 July 2018].

Juengst, E., 1998. What does *enhancement* mean? In: E. Parens, ed. *Enhancing human traits: Ethical and social implications.* Washington, DC: Georgetown University Press, pp. 29-47.

Kane, P.M., 2002. She offered herself up: The victim soul and victim spirituality in Catholicism. *Church History,* 71(1), pp. 80–119.

Kohrman, M., 2003. Why am I not disabled? Making state subjects, making statistics in post-Moa China. *Medical Anthropology Quarterly,* 17(1), pp. 5-24.

Kreeft, P., 1986. *Making sense out of suffering.* Cincinnati, OH: Servant Books.

Maguire, G.Q. and E.M., Mcgee., 1999. Implantable brain chips? Time for debate. *Hastings Centre Report,* 29(1), pp. 7-11.

Mahoney, J., 1989. *The making of moral theology: A study of the Roman Catholic tradition.* Oxford, UK: Clarendon Paperbacks.

Mccormick, R.A., 1984. *Health and medicine in the Catholic tradition.* New York, NY: Crossroad.

Merkel, R., G. Boer, J. Fegert, T. Galert, D. Hartmann, B. Nuttin, and S. Rosahl., 2007. Intervening in the brain: changing psyche and society. In: *Ethics of Science and Technology Assessment Volume 29.* Berlin: Springer Publishing.

Mitchell, C.B., E.D. Pellegrino, J.B. Elshtain, J.F. Kilner, and S.B. Rae., 2007. *Biotechnology and the human good.* Washington, DC: Georgetown University Press.

Parens, E., ed., 1998. *Enhancing human traits: Ethical and social implications.* Washington, DC: Georgetown University Press.

Paul VI., 1965. Pastoral constitution on the church in the modern world— *Gaudium et spes.* Available at: <http://www.vatican.va/archive/hist_councils/ii_vatican_council/documents/vat-ii_cons_19651207_gaudium-et-spes_en.html> [Accessed 6 July 2018].

Pius XI., 1930. Encyclical letter on Christian marriage— *Casti connubii.* Available at: <https://w2.vatican.va/content/pius-xi/en/encyclicals/documents/hf_p-xi_enc_19301231_casti-connubii.html> [Accessed 6 July 2018].

Pontifical Academy for Life., 2003. Concluding communiqué on the ethics of biomedical research for a Christian vision. Available at: <http://www.vatican.va/roman_curia/pontifical_academies/acdlife/documents/rc_pont-acd_life_doc_20030226_ix-gen-assembly-final_en.html> [Accessed 6 July 2018].

President's Council on Bioethics., 2003. *Beyond therapy: Biotechnology and the pursuit of happiness.* Available at: <http://www.bioethics.gov/reports/beyondtherapy> [Accessed 2 January 2009].

Quinn, K.P., 2000. Method in Catholic bioethics. *Kennedy Institute of Ethics Journal*, 10(4), pp. 353-363.

Reich, W.T. ed., 1978. *Encyclopedia of bioethics.* New York, NY: Macmillan Library Reference.

Reich, W.T., 2003. Shaping and mirroring the field: The encyclopedia of bioethics. In: J. Walter and E. Klein, eds. *The story of bioethics: From seminal works to contemporary explorations.* Washington, DC: Georgetown University Press, pp. 165-196.

Sabin, J. and N. Daniels., 1994. Determining "medical necessity" in health practice. *Hastings Center Report*, 24(6), pp. 5-13

Silvers, A., 1998. "Species-typical" functioning. In: E. Parens, ed. *Enhancing human traits: ethical and social implications.* Washington, DC: Georgetown University Press, pp. 95-123.

Stein, E., 1987. *Woman.* Translated by F.M. Oben. Washington, DC: ICS Publications.

Teilhard de Chardin, P., 1950. *L'energie spirituelle de la souffrance.* Paris, FR: Seuil.

United States Conference of Catholic Bishops., 1986. *Economic justice for all: pastoral letter on Catholic Social teaching and the U.S. economy.* Washington, DC: USCCB Publishing Services.

United States Conference of Catholic Bishops., 2009. *Ethical and religious directives for Catholic health care services.* 5th edition. Washington, DC: USCCB Publishing Services.

Vanier, J., 1998. *Becoming Human.* Toronto, ON: House of Anansi Press.

Veatch, R., 2003. Revisiting a theory of medial ethics: Main themes and anticipated changes. In: J. Walter and E. Klein, eds. *The story of bioethics: From seminal works to contemporary explorations*. Washington, DC: Georgetown University Press, pp. 67-91.

Waldby, C., 2002. Stem cells, tissue cultures and the production of biovalue. *Health*, 6(3), pp. 305-23.

Walter, J.K. and E.P. Klein., eds., 2003. *The story of bioethics: From seminal works to contemporary explorations*. Washington, DC: Georgetown University Press.

Walters, L. and J.G. Palmer., 1997. *The ethics of human gene therapy*. New York, NY: Oxford University Press.

World Health Organization., 2017. VMUN 2017 Background Guide. *The organ trade*. Vancouver: Model United Nations, The sixteenth annual conference. January 20-22, 2017.

Wright, G.R., 2000. Second thoughts: How human cloning can promote human dignity. *Valparaiso University Law Review*, 35(1), pp. 1-38.

8.

The harmony of metal and flesh: Cybernetic futures

Jacob Boss

Indiana University Bloomington

Abstract

Science fiction and philosophy are partners in providing certain fears found at the place where flesh meets metal. Thinking with embodied cyborgs reveals their dependencies, that is, specific examples of where the cyborg as aspirational figure can grow toward its liberatory potential. This growth is not into empty space, the liberatory cyborg presses against and ruptures networks of control woven from capital, ideology, law, and the many other systems of the world that have been part of this conversation. I recommend the use of the term petite cyborg to help us think about the vulnerable intermediate, a state that technological optimism and ecomodernism overlook, and a state that aspirational visions of the cybernetic future faithful to the cyborg as liberator cannot neglect.

Keywords: Cyborg, Cybernetics, Technology, Science Fiction, Ecology, Hybridity

> "Everywhere we remain unfree and chained to technology, whether we passionately affirm or deny it" – Martin Heidegger (1977, 4)

Introduction

I want to examine certain fears found at the place where flesh meets metal. The figure of the cyborg, the human-machine hybrid, aids in thinking through fears that through technological innovation we may fall prey to hubris, make ourselves monstrous, or make monsters that will destroy us and our world. Companion to such fears is the hope that we can develop a relationship with matter and technology that is not based on dangerous fantasies of domination and mastery. The fear and hope that swirl around the integration of technology with the human body, or the replication of human forms and functions by machines, can never be separated from the conditions that make developing, manufacturing, and distributing such technologies possible. As physicist Stephen Hawking (2016) warned, an automated society may bring comfort to all, or misery to most and excess to a few. Technology that can liberate may be used to bind more tightly.[1] Hawking (quoted in Cellan-Jones, 2015) has also used his prominence to warn that if our relationship to technology such as artificial intelligence is a competitive one, we may soon create our own destroyers. To explore the themes of relationship to technology, and technology's potential for enhancement and harm, I draw together real cybernetically enhanced humans, philosophers of technology such as Donna Haraway and Martin Heidegger, and works of apocalyptic and utopian fiction from Japanese director Hayao Miyazaki and Scottish author Iain M. Banks. Miyazaki's corpus is exemplary as ecologically concerned fiction, replete with tales of complex, interpenetrating and relational experiences with technology, and the devastation of the natural world by technologically enhanced humans and their engineered agents. Advanced technologies involving the infiltration, modification, and replication of the human person such as genetic modification, cybernetics, robotics, and artificial intelligence carry real risks, and exceptional promises, and the promises can obscure the risks when the painful intermediate stages of technological salvation are ignored.

I honor the anxieties expressed by scientists, philosophers, and authors of fantasy and science-fiction, by proposing a term to track the overlooked intermediate stage between our present moment and a future state of imagined technologically crafted salvation. I suggest the term *petite cyborg* for hybrids of living tissue and fabricated components with certain social and material dependencies.[2] By making the *petite cyborg* a distinct figure, I aim to draw attention to the network of dependencies that make its emergence possible.

Moving beyond well-cooked debates over whether or not any use of technology, or small modifications such as implantable RFID tags (such as are

---

[1] "If machines produce everything we need, the outcome will depend on how things are distributed. Everyone can enjoy a life of luxurious leisure if the machine-produced wealth is shared, or most people can end up miserably poor if the machine-owners successfully lobby against wealth redistribution. So far, the trend seems to be toward the second option, with technology driving ever-increasing inequality" (Hawking, 2016).

[2] A friend who qualifies as a cyborg according to my framing, brought to my attention that the designation of '*petite*' should be evaluated carefully for its potential to echo gendered denigration of women. In my use, it springs from the Marxist category of the *petite bourgeoisie*, a social stratum of certain dependencies, powers, and aspirations.

used to find lost pets) make us cyborgs or not, I propose that our present engagement with pacemakers, cell phones, smart homes, smart glasses, hip replacements, insulin pumps, sophisticated prosthetic limbs, etc. should be usefully understood as making us *petite cyborgs*. Some further marks of the *petite cyborg* include: Limited scale of cyberization, limited utility of cybernetic implants, dependence on capital, research, and power from national and international organizations such as the Defense Advanced Research Projects Agency. The *petite cyborg* is defined through dependency. Reliable access to electricity, proprietary (and often life-sustaining) parts, and surgical expertise, are all challenges being lived by cyborg-aspirants today. If the cyborg is to be more attractive than a goddess, as Donna Haraway, author of the influential *A Cyborg Manifesto*, once proposed it was, it is going to have to demonstrate its power to *liberate* from systemic dependencies and structural inequalities. The cyborg then should be understood as an aspirational ideal. The fully realized cyborg will possess the ability to sustain, augment, extend, and replace the components of their body independent of the global energy, information, and capital systems that *petite cyborgs* depend on.

Material dependencies are of course always tangled up with discourse and thinking with cybernetics should force a confrontation in society at large over viewing bodies as inadequate or deficient. The *petite cyborg* is a simulacrum in that it is a production imagined as a reproduction.[3] This matters because there is no original 'arm' or 'leg' that can be reproduced in a factory. Cybernetic limbs are simultaneously mass-produced (in the case of prosthetics) and personalized in the sense that they are attempts to replicate or extend components of individual human bodies. All *petite cyborgs* are thus haunted by the cyborgs they *could* be, even when enjoying their augmentations. Cyborg studies must be informed by the scholarship of queerness, disability, age, gender, race, colonialism, all the ways in which bodies and identities are conceived and controlled. *Petite cyborgs* face a double haunting of limited flesh and metal, and even a little progress might be taken away, or go awry. As scholar of feminist theology Sheila Briggs (2006) argues, along with the promise of technologically facilitated bodily transformation comes the "fear that the self might be perpetually trapped in hideous or tortured flesh" (157). The fate of a *petite cyborg* is to be always at the mercy of their benefactors. A prototype arm might be taken away, a charity might not be able to pay for the ongoing maintenance of a cybernetic leg. A military research outfit might demand that so many hours of the *petite cyborg*'s life be given over to testing and reporting on their implants. As

---

[3] See Baudrillard (1994) for more on how copies need no originals.

Haraway (1991) said, "the boundary between science fiction and social reality is an optical illusion" (149). Our bodies are changing to resemble the bodies of our imagined future, but the *petite cyborg* is not in control of the substance of their being. They are reliant on others to provide, fund, develop, install, repair, upgrade, and remove their cybernetic components. The current state of access to cybernetics is profoundly limited. A sophisticated cybernetic enhancement is beyond the ability of most humans to acquire due to both financial and clinical-discursive controls. Finding a way to liberate from limited access, as Haraway's better-than-divine cyborg must do, requires careful analysis of competing and overlapping visions for the cybernetic future.

In the following sections, I discuss how ecomodernists[4] such as Andrew Revkin and Ted Nordhaus promote a vision of a cybernetic future that is already upon us, without regard for the uneven global distribution of resources and research available in a world in which it is claimed that 'we' humanity, are now unified through technology. Political scientist Francois Gemenne's work on climate refugees and other bodies that vanish in the surging discourses of eco-politics problematizes the 'we' that Revkin delights in; I rely on Gemenne's observations as I explain in what ways the quest for a cybernetic future produces abject bodies or gives rise to a false sense of unity among humans.

To discuss ecology and cybernetics in science fiction, I turn to *Nausicaa of the Valley of Wind,* a graphic novel by Japanese director Hayao Miyazaki, whose animated works such as *Spirited Away* enjoy global acclaim. The titular character of *Nausicaa* struggles with machine technology, biologically altered humans, and a post-apocalyptic world haunted by the promise of extinction. My discussion of Nausicaa's ultimate rejection of mastering the world through technology is entwined with Bruno Latour's (2010) concept of "continuing to care for unwanted consequences" (26). Latour urges us to maintain contact with the unpredictable products of our technological creations. We may be enchanted or horrified by the consequences of what we create, but as Latour sees it, the greatest sin would be to turn away from, abandon, or reject those technologies. Instead, Latour advocates that though we cannot know where it will lead us, or into what messes we will have to deal with, we should maintain our parental relationship to our technology.

---

[4] The ecomodernist movement argues for what they term 'decoupling', where the environmentally deleterious results of human industry are separated, through advanced technical means, from impacts on nature.

The other vision of the future considered here is a utopian symbiosis of humanoids and machine intelligences. Scottish author Iain M. Banks invites readers to experience what he calls 'The Culture', where scarcity, disease, aging, infirmity, dementia, and arguably death have been conquered by human-like beings and their fellow citizens, artificial intelligences of varying sophistication. The Culture provides a touchstone for exploring ideas of community and relationality between technologically enhanced humanoids and machine intelligences.

As intermediate beings, *petite cyborgs* have limited ability to recreate, augment, replace, and enhance their bodies. Generally, they are greatly dependent on states and social welfare programs, on the generosity of corporate sponsors, on not-for-profit charities, and military research and development. They depend on national power infrastructure to charge their components; they depend on international manufacturing and distribution to acquire the materials with which their bodies are altered. *Petite cyborgs* are real people living with partially fabricated bodies; their dependencies point beyond themselves, to the promise and peril of independent cybernetic life. That is, a cyborg who wields a robot arm is -right now- also a cyborg who must recharge their body parts, is compelled to remove them and pay a power company so that their body can continue functioning. These limitations are invitations to overcome cybernetic dependencies.

Many of the figures that appear in the discussion that follows are concerned with the Anthropocene, the idea that our Holocene epoch has been replaced with one in which human beings have become a geologic force. For some theorists, the Anthropocene is predicated upon a never before seen enmeshment of human activity and global, geologic consequences. Itself a term for certain kinds of enmeshment, cybernetics broadly can refer to entanglement between living bodies and the nonliving substances of the earth. Investigating *petite cyborgs* allows us to see how those who think the climate crisis is an opportunity to reshape the world and our bodies, taking evolution both personal and planetary into their hands, gloss over who gets left out of grand visions of human planetary dominance.

Let us look at some examples of *petite cyborgs*, in order to better understand their limited ability to acquire and maintain their abiotic components, and how these individuals describe the impact of living cybernetic lives. These *petite cyborgs* demonstrate that integrating technology with the human body holds promises both social and somatic, and that some *petite cyborgs* are aware that achieving independence requires transforming the systems that allowed their emergence.

Living with cybernetics

In 2001 Jesse Sullivan was electrocuted so fiercely in a workplace accident that he lost both of his arms. Through the Rehabilitation Institute of Chicago, Sullivan was provided with a prosthetic arm responsive to his thoughts. Todd Kuiken, director of neuroengineering at the Center for Artificial Limbs at the Institute, describes Sullivan's prosthetic as a significant development in the quest to produce replacement limbs for injured soldiers. The Institute works in partnership with the National Institutes of Health and the Defense Advanced Research Projects Agency. Sullivan's cybernetics are notable for their ability to connect nerve clusters with a cybernetic limb and to provide sensations of pressure through the prosthetic. Sullivan is not a public figure, so little information is available about his cybernetic development since receiving the arm in the mid-2000s. Some of this may be due to Sullivan's desire to lead a private life, but I suggest that a significant factor that keeps Sullivan out of the public eye, compared to other *petite cyborgs*, is the aesthetic of his prosthetic. Sullivan's cybernetics are bulky and industrial in appearance, covering most of his upper torso and shoulders in plates, straps, wires, motors, sensors. The limbs themselves are plain, simple, functional; they are for doing, not exhibiting.

Nigel Ackland, who lost his arm in an industrial accident, is a public educator and amputee advocate. At WIRED Health in 2015, held at the Royal College of General Practitioners in London, Ackland gave a video recorded presentation on his experience. He began his journey into cybernetics when responsibility for his on-the-job injuries was disavowed by his employer. "I was told my company had really good insurance…a small piece of the insurance policy meant that instead of the private treatment I was told to expect I was dumped onto the National Health Service" (Ackland quoted in Wired UK Staff, 2015). Ackland's company used arcane insurance rules to avoid giving him superior care, forcing him to receive simple prosthetic devices from the public medical system. Ackland suffered from the incompatibility of his simple prosthetics with his body, resulting in pain, skin conditions, and limited limb function. The inadequacy of his prosthetics contributed to mental degradation, nightmares and rages. His family life and social life deteriorated, and he suffered heart problems. "This is what you get on the NHS after three or four years if you're lucky," he says in the video, holding up an 'arm' that looks like an industrial wrench (Ackland quoted in Wired UK Staff, 2015).

Ackland's turnaround came when he became a trial patient for RSL Steepler's bebionic prosthetic,[5] which is modeled on the human hand. The hand and its

---

[5] The bebionic hand was acquired from Steepler by Ottobock HealthCare in 2017.

mount work together to provide a replacement for Ackland's lost limb, a replacement designed to limit additional damage to his body. "This is the most important part of the whole thing," Ackland explains, "if you can't wear this every day all day without pain or discomfort it doesn't matter." Superior in dexterity and sensitivity to the NHS wrench-hand, Ackland's prosthetic hand is capable of delicate tasks such as cracking eggs and typing on a keyboard. "This is for shaking hands," he says, extending his hand, which appears metallic yet human, the fingers distinct, the joints articulated. "This is for being human." An article in *Wired* from May of 2015 reports that "from being stared at and laughed at, Ackland says he has now been accepted. He describes shaking hands with people using his bionic hand and seeing a genuine smile. Ackland reports that 'I see that smile as a sign of acceptance for who I am. Nobody ever asked to shake my hook'" (Artilce and Ackland quoted in Wired UK Staff, 2015).

What makes possible the availability of cutting-edge prosthetics at limb centers? "Foresight, empathy, balls," says Ackland, who also claims that "If you took the money the NHS uses to pay the Super-Managers you could provide 5,000 of these every year." Being one of the fortunate few to receive an advanced prosthetic, Ackland advocates for widespread adoption of similar technologies. He argues that cutting bloated administrative salaries would be a significant step toward making sophisticated prosthetics widely available. Ackland, who likens himself to a fortunate infant or toddler, equating his cyberization with rebirth, demonstrates how cyborgs are, as Haraway claimed, exceedingly unfaithful to their fathers. To move beyond the intermediate and dependent state of the *petite cyborg*, cyborg-aspirants must turn against the inequitable systems that birthed them.

Ackland makes the compelling argument that cybernetics can be a technology that restores us to humanity and restores our humanity to us. His case also demonstrates a core quality, and perhaps limitation, of what I count as a cyborg. Cybernetic limbs rely on interfacing with biotic components to control their movements. If a being were to have every scrap of flesh and tissue replaced with manufactured analogs, they would cease to be a cyborg as I have defined them, and would have become something else—intelligence supported by a machine substrate as imagined by transhumanists.[6] It may be that the only way to have autonomy is to cease being a cyborg.

Ackland attributes the lack of distribution of sophisticated cybernetic limbs to cost. The development, manufacturing, and distribution of cybernetics are

---

[6] See Rothblatt (2014), *Virtually Human* , Kurzweil (1990), *The Age of Intelligent Machines* (1990), and Moravec (1988), *Mind Children*.

totally dependent on capital, as is the *petite cyborg*. If the cyborg is to be a new socialist myth, as Haraway argues, it must be developed as a threat, not an aid, to capitalism. It must become subversive. The makers of cybernetics may be approaching the distribution of their devices cautiously, knowing that once cyborgs independent of the allocation of global energy and capital emerge from the ranks of *petite cyborgs*, their ability to capitalize on their human-machine dependents will evaporate.

Enthusiasm for a technologically innovative future that carries human beings into a new era of global wellbeing is espoused by ecomodernists, as exhibited in the collectively-authored *Ecomodernist Manifesto* (2015). In that document, ecomodernists present a picture of desirable development, great dense cities capable of supporting huge populations, vast fields of solar panels, even biodomes on the moon or Mars. Their naïve belief is that this techno-utopian bounty will be the inheritance of all humanity. Cybernetics is illustrative of the limitations in equitable funding and distribution faced by ecomodernist proposals. Sophisticated prosthetics are only available through great expense and investment in research. The authors of "Gender and Geoengineering" push back against the utopian vision of the ecomodernists, whose rosy portrait of the future skips over the intermediate phase between our current world and the future they imagine. "History provides countless episodes of environmental injustice where damage to the earth accompanies disproportionate injury to disempowered demographics. Environmental harms are rarely simply ecological concerns; they are also social and political concerns" (Buck et. al.,n 2014, 658).[7] The authors point out how harm following from climate change falls disproportionately on women, children, and the impoverished. There can be no such thing as a harmless transformation of the globe. Techno-utopian visions, whether of the ecomodernist sort or the cybernetic variety, often fail to take seriously the interim phase in which technologies of transformation are dependent upon uneven and unequal regimes and resources.

The authors of the *Manifesto* do not account for the massive suffering that seems inevitable in the middle period between the present, and the author's realizing of their vision for remaking the world. Gemenne has linked this fatal oversight to a false sense of the size of our in-group. In "The Anthropocene and Its Victims" Gemenne (2015) warns us that we blind ourselves to the plight of the disenfranchised with the pretend-capaciousness of the embrace of 'our' progress: "The Anthropocene [our new era of progress] might produce the false impression of a unified humanity" (173). To help the reader

[7] Holly Jean Buck, Andrea R. Gammon, and Christopher J. Preston (2014).

understand the potential for disproportionate impacts that industrialization makes possible, Gemenne argues that climate change-induced migration is a result of the release of greenhouse gases, which should be understood as a form of political oppression. What links this to our discussion of cyborgs is that this political oppression moves human bodies around by force and alters the conditions under which they may have access to cybernetics. The *petite cyborg* is not only at the mercy of capital, but also subject to the shifting landscape, to an environment stricken by human-driven climate change.

Topography of the noosphere

Let me provide a concrete example of the kind of global imaginary that Gemenne warns against. Andrew Revkin, an environmental journalist with ecomodernist sympathies, often highlights the importance of the Jesuit philosopher Pierre Teilhard de Chardin and his concept of the noosphere, which Revkin characterizes as global mind. Revkin connects the noosphere, an envelope of human thought or mind that is taken as enclosing the earth, to Darwin and his reading of the Darwinian concept of social sympathy. Revkin trades on the idea of evolution as progressive, insisting that Darwin saw differences between different human cultures as collapsible and believed? That the collapse of this distance is desirable and advantageous. The collapse of the barriers between human subjects is made possible, Revkin believes, through specific technologies that pierce an "artificial barrier" separating human beings from each other. Revkin (2014) claims that, thanks to communications technology, "we now have the ability to have a global *we*." Revkin (2014) identifies the most significant feature of the global mind as being that "an idea can move from here to Bangalore to Botswana right now. That's as important as the biogeophysical changes that are happening." In other words, Revkin believes that through technological means we are achieving Teilhard's vision. The global mind that Revkin describes is cybernetic in that it is composed of a latticework of fabricated strands and stations that convey data and translate that data into forms comprehensible to human users within that network. I challenge this conception of the 'global we', drawing on what I call the 'topography of the noosphere'.

The topography of the noosphere takes seriously the notion of global mind advanced by Revkin. How does Revkin's understanding of the global mind differ from Teilhard's? Teilhard was a Catholic theologian whose concept of a sphere of mind, while mediated through the scientific enterprise, was nevertheless grounded in a teleological process that found its resolution through union with a cosmic Christ. Revkin's global mind is a celebration of human technical inventiveness and industrial capacity. Revkin cites Teilhard in order to establish a foundation and lineage for his own concept of global mind but fails to explain *both* that the noosphere is a theological construct, *and* how a cybernetic world-mind can derive from Teilhard's noosphere, depending as it does on the existence of a cosmic Christ to orient its evolution. Nor does Revkin attempt to reconcile his use of Darwin as an extension of Teilhard's thought, despite the collapse of barriers described by Teilhard being entirely unlike the evolutionary processes Darwin described. Darwin understood social animals as capable of expanding the range of their empathy, while Teilhard's participants in the noosphere undergo a transformation and fusion of *soul*. This distinction is important, because Darwin's argument concerns social and biological transformation, while Teilhard's describes a metaphysical transformation.

Revkin's imagined noosphere is not some seamless, smooth globe of sacred energy, some Christ-aura filling the space evenly around the Earth. Revkin does not acknowledge that his noosphere is cratered like the surface of the moon, inscribed with deep pits and trenches where human access to information and technology are distributed unequally and unreliably. The participants in Revkin's 'we', the minds that rise and join with his vision of the noosphere, are those who possess sufficient language skills, technical knowledge, capital, energy, and equipment to plug in. As Kate Raworth (2014) reports in a critique of the Anthropocene concept, the production of science and technology remains largely created by men for men, most of them residing in North America and Europe. Revkin's noosphere similarly participates in the creation of an imagined 'we' that simply excludes much of humanity. The International Energy Agency (2015) reports that 1.2 Billion humans do not have access to electricity, and in 2015 *The Atlantic* reported that the majority of humans are not "active or regular" internet users. Revkin's noosphere, like the *petite cyborg*, is a construct of technology requiring enormous investments and research to produce and maintain and is entirely vulnerable to global systems of capital and energy. Hardly a model for liberation, and bearing little relation to Teilhard's cosmic Christ, Revkin's noosphere is fully enmeshed within the neoliberal economy. It serves as a chastening example of the kind of exclusionary 'inclusion' that Gemenne warns against. If we let our desire to build a bigger aggregate 'we' blind us to those who remain outside the circle we draw, we make ourselves responsible for their exclusion.

With the troubling limitations of this global imaginary in mind, I turn now to discussing two ways in which the technologically sculpted future of humanity has been imagined in science fiction. In what follows, I put the utopian idealism of the ecomodernists (along with Bruno Latour's concepts of stewardship and breakthrough) into conversation with these science fiction visions. One of the goals of this discussion will be to see whether and to what extent it is possible to allay the concerns about human relationships to technology as expressed by Martin Heidegger in *The Question Concerning Technology.* Heidegger worried that a fundamental connection between human beings who make things, the raw materials we use, the pure idea that is the template for a fabricated object, and the eventual product, has been lost. This relationship, according to Heidegger, once was characterized by a mutual appreciation and respect, expressed by an indebtedness between the components of made things.

Science fiction – imagining bodies into being

The perennial question "What can we entrust to the future?," runs through Japanese animator Hayao Miyazaki's opus *Nausicaa of the Valley of Wind,* a graphic novel thick with environmental themes. *Nausicaa* tells the story of a young ruler struggling to lead her people to survival as the remnants of post-apocalyptic humanity fight over the few remaining resources in the poisoned earth.

Nausicaa is named for the princess who aids Odysseus. Her name means 'burner of ships' and it is her fate to choose if she will aid or founder the vessel containing the hopes of the past. In *Nausicaa,* humans have changed their bodies to survive a world filled with deadly plants, air, and soil. Unaltered humans could not survive in this environment. In a chaotic swirl of hope and desperation the reader learns alongside the titular character that the poisoned land is cleansing itself, but the altered humans will not be able to survive in the new decontaminated world. The rebirth of the world means the end of humanity.

In the climactic final confrontation of the series, the technologically projected will of the past, in the form of a holographic image, appears to Nausicaa, demanding that she place her trust in the inventions and judgments of her ancestors. A single data archive survived the apocalypse, stewarded by an artificial intelligence that explains its nature and purpose to Nausicaa, who is now regarded as both king and messiah for her exploits. The keeper of the archive explains that it can fix the bodies of those who inhabit the ruined earth so that they will be able to survive the cleansing and reclamation of the land. Deep within the archive are vats from which a new, untainted humanity will spring, knowing only art and nothing of violence. Horrified, Nausicaa retorts, "Why didn't those men and women [the creators of the archive] realize that both purity and corruption are the very stuff of life?" (Miyazaki, 2000, 248) Her companion scoffs at the idea of humans who are ignorant of violence and know only art and peace, doubting that such creatures could count as human.

Each party in this confrontation sees the other as an impossibility. The voice of the past invalidates her existence and the existence of her fellow toxic humans. To the tenders of the techno-utopia to come they are inadmissible as human, their polluted bodies cannot enter into the coming paradise and live. This does not faze Nausicaa, who explains that "we have lived all these centuries with the Sea of Corruption. Extinction has long since become a part of our lives" (Miyazaki, 2000, 249). The keeper of the archives is aghast when Nausicaa rejects the future that the archive exists to ensure, branding her a nihilist for leaving the survival of humanity in the hands of nature, when technology can ensure its continuation. Nausicaa asserts that what matters is not the survival of any *kind* of life, but rather life itself, unshackled from a hubristic human belief in mankind's ability to control and direct it, to dictate survival and extinction. The term mankind is appropriate here, for the keeper of the archive appears through a projection of an older man, a scientist, explaining to the woman warrior-king-messiah that she must listen and obey him, or else humanity has no future.

Philosopher Bruno Latour has advanced a concept of 'breakthrough' in the relationship between humans and nature—or perhaps, in the dissolving of the boundaries between human and nature—that finds a devastatingly apt example in Nausicaa, where to continue to care is to shepherd nature on through to the destruction of the human species.[8] "But what is the breakthrough itself then?," Latour (2010) asks (26). "If I am right," he further argues, "the breakthrough involves no longer seeing a contradiction between the spirit of emancipation and its catastrophic outcomes but accepting it as the normal duty of continuing to care for unwanted consequences, even if this means going further and further down into the imbroglios" (Latour, 2010, 26). How deep can these imbroglios go, if we resolve to continue to care for the consequences of our technology? *Nausicaa* provides the reader with no reasonable basis on which to expect that humanity will survive the purification of the earth. Nausicaa prefers to allow the planet to determine if human beings will be part of its future, rather than surviving through a process that imposes separation between humans and nature, or that establishes humans over nature as its masters.

What Latour is advocating with respect to the relationship between technology and humanity through his rereading of Frankenstein's monster in his essay *Love Your Monsters,* is applicable to the example of *Nausicaa.* For Latour (2010) the monster represents our technological achievements

---

[8] Latour references breakthrough as an allusion to the Breakthrough Institute, an ecomodernist think tank.

disavowed, the unanticipated products of research we are embarrassed to be seen with. "Dr. Frankenstein's crime," Latour (2010) argues, "was not that he invented a creature through some combination of hubris and high technology, but rather that he abandoned the creature to itself" (21). Latour charges us with the sin of abandoning our spraying machines and gene therapies, our radiation consoles and selective breeding programs, all designed to seize control of the systems of the Earth. He proposes that abandonment is not the answer. Loving our monstrous creations may provide the opportunity to turn them to good ends. "Our sin is not that we created technologies but that we failed to love and care for them" (Latour, 2010, 22). Nausicaa wrestles with technology that has seized the role of father and master for humanity, while Latour suggests the image of humans as parents to technology who can either neglect or love their offspring.

I find Latour's argument that we must continue to care for the unanticipated consequences of our technologies appealing, but he does not account for the dangers in human-object relations as described by environmental historian Tim Lecain. Lecain (2015) explains the neo-materialist perspective on the relationship between humans and things as "partnerships humans formed with powerful material things whose potentialities often pushed them in directions they neither envisioned nor intended" (20). Latour claims that we must both love technology instead of abandoning it once it is created and be willing to care for the unwanted consequences of our actions and interactions. We might say that Nausicaa's solution to loving something that leads those that interact with it into a sense of division and mastery is to love it to death. She harnesses the technology of the ancient world in order to put an end to the same. "All things are born from darkness and all things return to darkness," she tells the archive, "and now it is time for you to return to darkness" (Miyazaki, 2000, 250). In the end she breaks with technology that would allow control over the future, choosing to live alongside the bioengineered flora and fauna that will eventually make the earth uninhabitable for humans. Latour (2010) speaks approvingly of being willing to follow along with unintended consequences, to 'roll with it' instead of trying to arrest the unfolding of events. Where it seems Latour must break from Nausicaa is when she abandons mastery through technology, preferring to bury the devices forever. The kink or confusion in this tableau as it relates to Latour's thinking is that Latour does not want humanity to abandon its technologies of mastering nature. But if we accept Lecain's argument, it appears we are technology's *followers*, not its masters, tricked like the people of Nausicaa's wasted earth into believing we are the architects of our future. Nausicaa exposes and occupies space beyond the possibilities that Latour describes. Latour's commitment to technologies of mastery stem, I speculate, from his enduring fascination with the work of political theorist and Nazi jurist Carl Schmitt, and the significance of Schmitt's friend/enemy distinction to Latour's environmental project, where there are friends of the earth sciences to harness, and enemies to overcome.

To render Nausicaa's struggle in cybernetic terms, ultimately the entanglement of sophisticated technology and biology becomes unbearable for the people of the earth. They feel that whatever mastery technology once provided has been replaced with slavery to technology. They have become what Heidegger (1977) calls the "standing reserve" (8). The standing reserve is an image and a rubric. It is the image of the storage room, which expresses the stasis of being on call, subject at every moment to the possibility that you will be required to fulfill your responsibilities in accordance with your capacities. It is a rubric, in that by thinking through it, we are led to determine if we view any given thing as something we already understand, in its potential, form, and function. If we are fully convicted in our knowledge of a thing, what potential of action, as well as relation, has been foreclosed? In the standing reserve, held ever ready and ever bound, there is no possibility of a spontaneous unfolding. In Nausicaa's world, human beings who serve the Ark live lives arranged by technology; they are servants to a future they are physically incapable of inhabiting. True to her name, Nausicaa burns the ship carrying humanity inexorably forward. Where they end up, only the tides will tell, receiving them into the grave or delivering them to new shores.

If Nausicaa serves as a portrait of humanity captured within the standing-reserve, it is also a promise of human liberation from the prison of technologically imposed stasis. Nausicaa does not derail the plans of the technocrats in order to usher in an era of eco-primitivism, she does not replace one utopian vision with another. It is not that Nausicaa's people find freedom by living without technology, but that they are ultimately delivered into a world where technology is not associated with mastery over the earth and the future of life; a place where new possibilities have room to emerge. Here I return to 'The Culture', the space-faring civilization created by Scottish science fiction author Iain M. Banks, to compare the relationship to technology that Banks conceives of with that found in *Nausicaa.* In its optimism, technological mastery of matter, and sentient AI, The Culture resembles the future sought by transhumanists such as Rothblatt and Moravec, Culture citizens appear as uninhibited as imagination allows. In The Culture as described by Banks, humanoid and AI citizens live beyond scarcity, they do whatever they want. Citizens live on vast vessels that travel between stars, and in the tens of thousands of years of The Culture's development it has become unfashionable to live on worlds capable of supporting life. Banks (1994) explains that The Culture disdains to live on worlds because, "The Culture, as is made clear in *Use of Weapons,* regards terraforming generally as ecologically unsound; the wilderness should be left as it is, when it is so easy to build paradise in space from so little." Consequently, AI-run manufactured ring-worlds and vast city-ships host most of the population.

There are hints even within Banks's work that nearly limitless power may not provide enduring satisfaction. On one of these manufactured worlds, in *The Player of Games*, two Culture citizens debate the joy possible in a life where everything is within reach. They play a game, which one enjoys and the other disdains. "Don't you get *any* pleasure from this sort of thing?" Asks the one. "No," says the other. "And neither will you, after a while." The first citizen shrugs, "Till then, then" (Banks, 1988, 3).

As a vision of a cybernetic future, it is unexcelled in its optimism and promise of what humans and machines can accomplish together. Citizens of The Culture, fleshy and machine, enjoy godlike powers to remake their bodies and environments. They are free from disease, have conquered aging, and can recover from almost any injury. If they die, they can restore themselves from a backed-up copy. Though they possess great power, their ability to instrumentalize all matter cuts them off from the possibility of new relations with matter. Heidegger (1977) anticipates the prevalence of this state of affairs, seeing it in his own time: "silver is that out of which the silver chalice is made. As this matter (*hyle*), it is co-responsible for the chalice. The chalice is indebted to, i.e. owes thanks to, the silver for that out of which it contains…at the same time indebted to the aspect (eidos) of chaliceness" (3). It is useful to the aspiring cyborg to see this as a description of a positive affective commitment between those who make objects, the substance the object is made out of, and the idea that guides the shaping of the object. When this has been broken, it is to our detriment. When our conception of objects and substances is restricted to certain views, and to their instrumental value, we have cut ourselves off from the ways in which they might surprise us with their potential. Moreover, failing to recognize that material, concept, and method are interdependent leads to an overconfidence that rests on the power to impose shapes on the universe, instead of a respectful commitment to the constituents of all shaping. As an aspirational ideal, the cyborg is aware of its substance, materially, ideologically, sociologically, and does not mistake reshaping the world as a fundamentally different practice than reshaping its own body.

The enfleshed and machine citizens of The Culture share respectful and grateful relations with each other, but as intelligent beings the machine lifeforms of The Culture no longer belong to the categories of technology with which Heidegger is concerned. What The Culture achieves is power, which can appear liberatory insofar as its wielders can dictate many of the conditions of their lives. By instrumentalizing all matter, The Culture makes it impossible for objects to, in Heidegger's terms, reveal themselves. If the makers of things, whether human or AI, were not oriented to the world in terms of mastery and utility, then they could discover that objects have the capacity to reveal themselves to the respectful partner. The worldview that operationalizes all matter, and leads to standardized manufacturing through technologies of mastery, denies objects the opportunity to reveal themselves. If objects are not allowed to reveal themselves, we remain trapped in impoverished worldviews and relational possibilities because by lauding ourselves as the technological masters of manufacturing we have lost the perception of indebtedness that positions us respectfully toward matter. With the loss of this indebtedness, Heidegger argues, our creative capacities are dammed up. Consequently, just as standardized forms are imposed on matter, so too on relations, where we come to regard each other only in terms of standardized utility.

Toward free relations

Thinking with embodied cyborgs reveals their dependencies, that is, specific examples of where the cyborg as aspirational figure can grow toward its liberatory potential. This growth is not into empty space, the liberatory cyborg presses against and ruptures networks of control woven from capital, ideology, law, and the many other systems of the world that have been part of this conversation. I recommend the use of the term *petite cyborg* to help us think about the vulnerable intermediate, a state that technological optimism and ecomodernism overlook, and a state that aspirational visions of the cybernetic future faithful to the cyborg as liberator cannot neglect. We captive beings have been augmented, infiltrated, enhanced, extended, or healed, and we remain unfree. Caring for the vulnerable intermediate requires proceeding in our cybernetic growth with the utmost consideration. Respect, and so consideration, can, as we've seen from Heidegger, can arise in the recognition of chains of dependency. Latour too recognizes the cybernetic world we live in, correctly observing that we are constantly engaged in the mutual unbinding and binding up of each other. [9] But recognition does not necessarily lead to respect or care, or partnership.

Does Latour invite partnership? Following from Schmitt he must at least extend an invitation among friends. As difficult as it is to pin Latour down, his

---

[9] Latour (2010) writes: "Every day in our newspapers we read about more entanglements of all those things that were once imagined to be separable — science, morality, religion, law, technology, finance, and politics. But these things are tangled up together everywhere" (23).

positions, though always colored with mastery, at least contain the possibility of partnership. We either (wrongly) seize control, or we stay with the trouble, allowing events to unfurl, committing to following the consequences of our attempts to enact mastery, though we may finally understand that mastery or influence of any kind is always interdependent and hybrid. Acknowledgment of hybridity opens the door to partnership between the human and nonhuman.

The mastery that is possible, once independent mastery is rejected, is one that is shared by interdependent beings. My reading of The Culture shows that extreme forms of technical mastery do not, in themselves, cultivate ethical orientations or liberate from narrow ways of thinking and conceiving. Nausicaa's story demonstrates the hubris and oppression that can arise from projects of technologically facilitated salvation.

Nausicaa and The Culture are both stories that involve negotiating human-machine relationships. Turning to nonhumans helps us think through possibilities of partnership and indebtedness beyond exclusively human relations. How would the thinkers I have engaged with include nonhumans in the pursuit of emancipation? An enduring partnership that acknowledges mutual indebtedness seems unlikely to emerge from Lecain's view. For Lecain (2015), substances and objects act on and shape human culture and experience; humanity is a product of matter. Lecain at times describes this relationship as a partnership, but claims that a substance like coal, for example, "shaped the humans who used it far more than humans shaped coal" (22). Lecain (2015) concludes that we "may have to undergo a very difficult process of disentangling ourselves from some very powerful material things that have increasingly come to dictate our collective fate" (22). Nausicaa would agree with Lecain on this point, while Banks provides a sympathetic view of human-AI partnership.

The cyborg ideal, liberating from structural and systemic inequities, exists in virtual worlds, and is embodied in us. Haraway called the barrier between the virtual and our lived social reality an optical illusion; the virtual is part of the Heideggerian chain of indebtedness, we dream with and through it. We are all *petite cyborgs*. There is no escaping technology. Plastics settle to the deepest parts of the ocean and nestle in our vital organs. Without iron in our blood, we could not live. Frankenstein's compounded creature is not alone, it is us. Recognizing we are patchwork cyborg beings from the start gives us certain opportunities to grapple with the challenges we envision in the pursuit of the cyborg ideal. The most significant challenges in manifesting our cyborg future are not, I believe, the challenges of technological inventiveness, but rather those posed by heartless capitalism, misguided yearnings for an imagined natural state for bodies, and a hubristic faith in our mastery of the material of the universe. Revolution against socio-technological tyranny requires clear-eyed study of the promise and the threat that the *petite cyborg* represents. This entails allowing the *petite cyborg* to direct our analysis of systemic inequalities. As technologies are endlessly unfurled to the mingled cries of excitement and horror rising from *petite cyborgs*, we should understand both reactions as appropriate. In our pursuits we must not be so arrogant as to believe that humans can independently master the universe. or allow ourselves to be petrified into inaction by fears of undermining some essential quality of humanity. In the harmony of metal and flesh lies both the possibility of continued bondage and future freedoms.

References

Banks, I. M., 1988. *The player of games.* New York: Orbit.

Banks, I.M. A few notes on The Culture [online]. Available at: <http://www.vavatch.co.uk/books/banks/cultnote.htm> [Accessed 7 June 2018].

Baudrillard, J., 1994. *Simulacra and simulation.* Translated by S.H. Glaser. Ann Arbor: The University of Michigan Press.

Bennett, J., 2010. *Vibrant matter.* Durham: Duke University Press.

Briggs, S., 2006. Digital bodies and the transformation of the flesh. In: V. Burrus and C. Keller, eds. *Toward a theology of eros.* New York: Fordham University Press. pp. 153-166.

Buck, H. J. et al., 2014. Gender and geoengineering. *Hypatia,* 29(3). pp. 651-669.

Cellan-Jones, R., 2014. Stephen Hawking warns artificial intelligence could end mankind. *BBC News* [online]. Available at: <http://www.bbc.com/news/technology-30290540> [Accessed 12 June 2017].

Chandler, A., 2015. How much of the world has regular internet access? *The Atlantic* [online] Available at: <http://www.theatlantic.com/technology/archive/2015/09/united-nations-internet-access/406552/> [Accessed 3 December 2017].

Durkheim, E., 1912. *The elementary forms of religious life.* Translated by K.E. Fields, 1995. New York: The Free Press.

Gemenne, F., 2015. The anthropocene and its victims. In: C. Hamilton, F. Bonneuil, and C. Gemenne, eds. *The anthropocene and the global environmental crisis.* New York: Routledge, pp. 169-174.

Haraway, D. J., 1991. *Simians, cyborgs, and women: the reinvention of nature.* New York: Routledge.

Hawking, S., 2016. *Stephen Hawking AMA.* Reddit [online]. Available at: <https://www.reddit.com/r/science/comments/3nyn5i/science_ama_series_stephen_hawking_ama_answers/cvsdmkv/> [Accessed 3 March 2018].

Heidegger, M., 1977. *The question concerning technology, and other essays.* Translated by W. Lovitt. New York: Harper & Row.

International Energy Agency., 2015. World energy outlook. Available at: <http://www.worldenergyoutlook.org/resources/energydevelopment/> [Accessed 14 December 2015].

Latour, Bruno., 2010. Love your monsters. *Breakthrough Journal*, 2 [online]. Available at: <https://thebreakthrough.org/index.php/journal/past-issues/issue-2/love-your-monsters> [Accessed 3 March 2018].

Lecain, T.J., 2015. Against the Anthropocene: A neo-materialist perspective. *International Journal for History, Culture and Modernity.* 3(1), pp.1–28.

Miyazaki, H., 2000. *Nausicaa of the valley of wind.* Vol. 4. Viz Communications.

Raworth, K., 2014. Must the anthropocene be a manthropocene? *The Guardian* [online]. Available at: <http://www.theguardian.com/commentisfree/2014/oct/20/anthropocene-working-group-science-gender-bias> [Accessed 12 December 2017].

Revkin A., 2014. Paths to a 'good' Anthropocene. *AESS 2014 Conference in NYC* [online]. Available at: https://youtu.be/oiheqlkgd3q [Accessed 2 February 2018].

Sententiaeantiquae., 2015. Was nausikaa a "ship-burner"? Speaking names and etymology. Available at: https://sententiaeantiquae.com/2015/07/30/was-nausikaa-a-ship-burner-speaking-names-and-etymology/ [Accessed 3 April 2018].

SWNS TV., 2012. "Terminator" arm is world's most advanced prosthetic limb. Available at: <https://youtu.be/_qupnnroxvy> [Accessed 1 February 2018].

Temperton, J., 2015. The world's most advanced prosthetic is changing lives. *Wired* [online] Available at: <http://www.wired.co.uk/news/archive/2015-04/24/nigel-ackland-prosthetics-pioneer-wired-health-2015> [Accessed 12-14-16].

Wired UK Staff., 2015. Nigel Ackland: The world's most advanced prosthetic arm. Available at: < https://www.youtube.com/watch?v=Mrx1e9wi3NM > [Accessed 1 February 2018].

9.
Embodiment matters: integral ecology, science, the promises and challenges of radical life extension, and socio-ecological flourishing

Christopher Hrynkow
University of Saskatchewan

Abstract

The chapter brings Pope Francis' teachings on integral ecology into dialogue with the aims and scopes of the scientific project. It does so in order to consider the ethics and implications of radical life extension, with a particular focus on the relevant impacts upon socio-ecological flourishing. More specifically, this chapter forms its argument by highlighting the relational framework offered by Pope Francis in *Laudato Si'*. It then applies Francis' articulation of integral ecology to issues that can be mapped at the junctions of science, social justice, and the common good as they relate to human enhancement. The results of this approach to the subject matter problematize some of the manifestations of elitism active in human enhancement, particularly as those tensions relate to radical life extension.

Keywords: Pope Francis, Integral Ecology, Socio-Ecological Flourishing, Radical Life Extension, Human Enhancement.

Introduction: "nothing is itself without everything else"

> We are drawn out of the safe pace of our life. We are called to help others to lead an authentic life, which means a life that is always one that serves others. In the end, living authentically is a question of interdependence, of committing one's self to the common good. It is, in its most basic reality, a question of a person coming to the understanding that one's good, that your good, is intimately tied to the

> good of everyone else. And that an authentic life is one that embraces a responsibility to the society in which one is a very, very small part (Chipman, 2017, 8).

This Ignatian take on Pope Francis' teachings can be read as situating authentic human roles within a realm of interdependence and social responsibility that is properly oriented towards the common good. Within that framing, duties to promote the common good extend to Catholics, secular scientists, and scientists who are Catholics. Such a conclusion draws energy from the premise that "everything is integral and interacts with everything else. This means that nothing is itself without everything else. There is a commonality, an integrity, an intimacy of the universe with itself" (Berry, 2002, 43). Within his articulation of moral imperatives to care for our common home, Francis makes the connections implied and envisioned in the concept of socio-ecological flourishing as unfolded and employed below. This confluence is indicative of the fertile ground for buttressing cultures of encounter and dialogue that the current pope's green orthopraxis provides. That potential is something that has been evident, though certainly not completely realized, in the life of Catholic communities struggling to model social equality and ecological sustainability, since Jorge Mario Bergoglio became Bishop of Rome in 2013. At that time, he chose the regal name Francis. This choice signaled his intensions for a papacy that would deal with many issues, inclusive of the proper role for science in the contemporary world, within a rubric coupling concerns for substantive peace, social justice, and vibrant ecology.[1]

The present chapter will explore particular ways that socio-ecological flourishing can be motivated and deepened by the teachings of the current Bishop of Rome. It will approach that task with reference to issues that lie at the intersections of science, religion, and society. More specifically, this chapter's reading of Pope Francis's articulation of integral ecology will be employed to explore reasons to both caution against and welcome recent developments in human enhancement that are aimed towards radical life extension.

In brief, integral ecology understands humans to exist in intertwined social, ecological, and cosmological relationships. Working from that perspective on deep diversity, this chapter will navigate tensions and promises brought into being by both currently available human enhancements and aspirational technocentric framings of radical life extension. The latter will be the primary

---

[1] On the transformative value of the coupling of social justice, substantive peace, and ecological health see Hrynkow (2014).

focus. Herein, a particular concern will be approaches to radical life extension that would take people beyond an ecospiritual framing of the essential nature of the human person as embodied and situated within a diverse world, which is properly marked by socio-ecological flourishing.

To achieve its goals, this chapter constructs a theo-ecoethical anthropology that is centered upon the essential nature and situation of the human as an embodied person-in-relationship with other humans, the more-than-human members of the Earth community,[2] and the rest of the cosmos. It will then employ that anthropology as a lens to survey a selection of current and proposed practices of human enhancements. While Pope Francis' articulation of integral ecology provides the primary theoretical lens, dialogue partners for this chapter will also include advocates for both ecospiritual worldviews and human enhancement drawn from amongst academics, activists, and practitioners. This eco-methodological approach to the subject matter brings into focus some advantages and disadvantages of human enhancements, which come into view in light of a theo-ecoethical perspective on what marks us a human. In doing so, it also provides a demarcation range past which human enhancements are deemed to be damaging to human persons-in-relationship when they are contemplated in light of an ecospiritual anthropology. That range, and the problematic implications of seeking to transcend it, are articulated for the consideration of the reader. Through these means, the reader is provided with an eco-ethical critique of transhumanism that brings a perspective valuing diversity at the margins into conversation with advocates of radical life extension. This critique is cogent, as will become evident below, because advocates for radical life extension tend to represent a constituency with a great deal of cultural and economic capital, placing them in a more central position with 'power over' relatively more peripheral social and ecological others.

Pope Francis on rapidification and the relational framework of integral ecology

Admittedly, not all Catholics are aware of the authoritative nature of papal encyclicals, which are held to binding on the conscience by moral theologians working within their faith traditions (see Gaillardetz, 1997; 2003). Yet, many

---

[2] This chapter employs the term 'more-than-human' in line with its use in ecoethical contexts to connote members of the presently existing Earth community who are not human. The term is employed to accord respect to such beings. 'More-than-human' also serves as something of corrective to language use that has the effect of tacitly reinforcing anthropocentric worldviews. For perhaps the most influential use of 'more-than-human' see (Abram, 1996/1997).

Catholics are aware that the Pope released an important document on caring for our common home in 2015. In that encyclical letter, *Laudato Si'*, Francis writes:

> The continued acceleration of changes affecting humanity and the planet is coupled today with a more intensified pace of life and work which might be called "rapidification." Although change is part of the working of complex systems, the speed with which human activity has developed contrasts with the naturally slow pace of biological evolution. Moreover, the goals of this rapid and constant change are not necessarily geared to the common good or to integral and sustainable human development. Change is something desirable, yet it becomes a source of anxiety when it causes harm to the world and to the quality of life of much of humanity (Francis, 2015, #18).

To consider the consequences of breaking the above-referenced 'pace of biological evolution' as it has operated on our planet, this chapter elects to begin its argument by highlighting its take on the relational framework offered by Pope Francis in *Laudato Si'*. More specifically, the present chapter applies Francis' teaching on integral ecology to issues that can be mapped at the junctions of science, social justice, and the common good as they relate to human enhancement. Here, it is paramount to claim to a certain understanding of ecology.

In a significant sense, ecological ethics mimic a feature of the scientific field of ecology, which in part stakes out its territory as the examination of whole systems. This approach causes tensions with other scholarly approaches in part because academics working in fields as diverse as politics and epidemiology speak of "the ecological fallacy" as a shorthand for the mistake of collapsing qualities properly belonging to a group into those attached to an individual (see Piantadosi, Byar, and Green 1988; Knopp et al. 2017). They also write critically of the 'naturalistic fallacy', the supposed error of deriving an 'ought' from an 'is', that is prominent in the academic study of philosophy (for a positive take on the concept and its ethical application from a trained biologist see Weber, 2016). Taken to an extreme, such labeling implies that ecological studies are almost automatically invalid (Wills and Soneji, 2018) as are approaches like biomimicry, which depend upon natural structures as sources of inspiration to help address complex problems facing people (see Dicks, 2016). However, avoiding ecological thinking and other approaches that engage the Earth community as sources of wisdom and guidance for human action misses a potentially transformative feature of the field of ecology. That transformative feature is brought forward by several green activist academics (see, for example, Krieger, 2015; O'Sullivan 1999; 2005) and Earth community-embracing ethicists of the theological and philosophical variety (see, for example, Hogue, 2010). It can be summarized in the premise that ecology is about entities as they exist in relationship (see Rasmussen, 1996; Eaton 2013; Benzoni, 2007). In theo-ethical terms, it is thus hard to extract an anthropology that fails to see ourselves as existing within human-Earth-divine relationships (see Hrynkow, 2016).

Amongst theists, as affirmed by Francis (2015) in *Laudato Si'*, that essential location implies duties to self, neighbor, the rest of creation, and God understood as existing with a web of potentially mutuality-enhancing relationships. This worldview is evident in many features of Francis' green orthopraxis, including by the Pope's co-occurring use of terms like integral ecology, interdependence, justice, and the common good. 'Integral ecology' as expressed in the encyclical, has genealogical roots dating back to the Franciscan Liberation theologian Leonardo Boff's (1995) earlier use of the term. Further, drawing upon a scientific progeny, the Pope Francis: places integral ecology as inherently social, connecting the cry of people living in poverty and otherwise on the margins with the cry of Mother Earth so that "today…we have to realize that a true ecological approach *always* becomes a social approach; it must integrate questions of justice in debates on the environment, so as to hear *both the cry of the earth and the cry of the poor*" (Francis, 2015, #49. Emphasis in the original).[3]

Integral ecology as articulated by the present Bishop of Rome thus couples the personal and political, along with the social and spiritual, understanding them as part of an interconnected whole. One way to summarize the aspired-for-outcome implied in *Laudato Si'* is to speak of support offered by Francis (2015) for socio-ecological flourishing.[4] In this regard, he supports a participatory socio-ecological ethics and an integral way of being for this world. When viewed in light of the of goal of fostering socio-ecological flourishing, the fate of the human and larger Earth communities are thus seen as intertwined on a deeply personal level that feels ecological ruin and social injustice as suffering in our own bodies with transformative implications for our communities, social structures, and political lives (see the conclusion to this chapter). In sum, the ecospiritual anthropology offered by Francis (2013; 2015) is one that views us as created beings in embodied relationships within a created cosmos.

---

[3] For more on this connection between the social and ecological in Francis' thought see Hrynkow (2019).

[4] On the cogency of socio-ecological flourishing see Hrynkow (2017).

Applying integral ecology to the scientific project

As may already be evident from that brief situating of the term in the preceding section, the scientific project writ large offers support for integral ecology. Yet, modernist science has also underpinned many efforts that have reduced the capacity for social-ecological flourishing in this world. The presence of such a mixture of tension and promise in the scientific project is a central theme in *Laudato Si'*, one that this chapter returns to in its conclusion. Francis (2015) wades into these complicated and turbulent, but at the same time potentially life-giving, waters in order to both critique modernist science and show how a more integral science does offer a path to help humanity exit from the malaise of intertwined social and ecological crises.

In terms of his critique, Francis considers how what are in reality inseparable human-Earth-divine relationships are disrupted by the worldview he names as a "technocratic paradigm" (#111), which has been buttressed by modernist science's tendency toward segmentation—in the simplest terms, that is, breaking things apart in order to study them without ever putting them back together as wholes.[5] Reconstituting something thus segmented is, of course, currently impossible in case of procedures like vivisection but the insight in play here also refers to a tendency to fail to put scientific facts back into the larger socio-political, ecological, planetary, and cosmological contexts. For Francis (2015), such narrowing of horizons comes with social and spiritual implications.[6]

Adding momentum to this segmentation are presently normative notions of progress, which are far too narrow in their accounting of social and ecological externalities and tend towards offering what Thomas Berry (2006) names as a saccharine 'WonderWorld' that is, in reality, manufacturing a 'WasteWorld'. Specifically, Francis (2015) shades such a limiting manifestation of progress as driven by the aforementioned technocratic paradigm that amasses undue power, which does not respect social and ecological limits. He critiques this narrow conception of progress in light of the proper "place of human beings and of human action in the world" (Francis, 2015, #101). For Francis, the marriage of scientific method and technocratic power has been particularly problematic:

> The basic problem ... is the way that humanity has taken up technology and its development *according to an undifferentiated and one-dimensional paradigm*. This paradigm exalts the concept of a subject

[5] On this framing of the integrity of knowledge compare Lonergan (1957/1992).

[6] On the use of 'horizon' here compare Lonergan (1972/2017).

> who, using logical and rational procedures, progressively approaches and gains control over an external object. This subject makes every effort to establish the scientific and experimental method, which in itself is already a technique of possession, mastery... This has made it easy to accept the idea of infinite or unlimited growth, which proves so attractive to economists, financiers and experts in technology. It is based on the lie that there is an infinite supply of the earth's goods, and this leads to the planet being squeezed dry beyond every limit (Francis, 2015, #106).

Notwithstanding these concerns, Francis (2015), who is trained as a chemist himself, is emphatic that science helps to bring into a view the nature of this disruption as it aids in such tasks as diagnosing anthropogenic climate change or discerning the effects of toxicants on human health. Hence, it can be concluded that Francis (2015) sees great promise in science when it is mediated by the concerns of integral ecology. Francis (2015) even extends value to technological-oriented science in one of the famous passages from *Laudato Si'*. That teaching is indicative of his pastoral approach that invites all people into dialogue about the fate of our common home. This is an approach which may frustrate eco-activists for being too accommodationist, even as it employs scientific language to ground some of the Pope's more esoteric reflections. Nonetheless, these threads all relate back to his articulation of a theo-ecoethical anthropology:

> Technoscience, when well directed, can produce important means of improving the quality of human life, from useful domestic appliances to great transportation systems, bridges, buildings and public spaces. It can also produce art and enable men and women immersed in the material world to "leap" into the world of beauty. Who can deny the beauty of an aircraft or a skyscraper? Valuable works of art and music now make use of new technologies. So, in the beauty intended by the one who uses new technical instruments and in the contemplation of such beauty, a quantum leap occurs, resulting in a fulfillment which is uniquely human (Francis, 2015, #105).

This quote is further representative of how Francis thirsts for responses to social and ecological crises that both recognize moral concerns for all of humanity and are rooted in a deep understanding of our anthropological location within an interdependent world. Moreover, he shades these paths as providing spaces for transformative action,[7] when, for example, technoscience is humanized and properly directed. Below, this chapter returns to that connection between human nature and interdependence within the Earth community understood as providing a basis for a theo-ecoethical anthropological lens that can be suitably employed to evaluate radical life extension projects. Here, it is important to note how Francis is upholding an imperative for the humanization of science and progress, which is certainly not neo-Luddite in the sense of being anti-technoscience (see Roszak, 1986/1994). It is in that light that of a qualified embrace of scientific insights and results that Francis' characterization of technology's fusion with consumerism as too often serving myopic and segmented interests is perhaps best read. In this regard, *Laudato Si'*, while offering the incarnation of the principles of integral ecology as antidote, questions the limited conception of both human freedom and quality of life that a 'techno-economic paradigm' spawns. Given meaning to that term, Francis (2015) teaches, "this paradigm leads people to believe that they are free as long as they have the supposed freedom to consume...We have too many means and only a few insubstantial ends" (#203).

Integral ecology as a method for science and politics to better foster socio-ecological flourishing

That lack of a substantive *telos* has negative consequences for the entire Earth community. As Francis tellingly writes, "a sober look at our world shows that the degree of human intervention, often in the service of business interests and consumerism, is actually making our earth less rich and beautiful, ever more limited and grey, even as technological advances and consumer goods continue to abound limitlessly" (Francis, 2015, #34). Thus, in general theo-ecothical terms there emerges a need for the scientific project to integrate the precautionary principle as an enabling mechanism that helps ensure humanized applications of technoscience. Indeed, Francis explicitly endorses this principle and its associated methodology for making decisions because of the protection it affords to beings on the margins:

> This precautionary principle makes it possible to protect those who are most vulnerable and whose ability to defend their interests and to assemble incontrovertible evidence is limited. If objective information suggests that serious and irreversible damage may result, a project should be halted or modified, even in the absence of indisputable

[7] On social, ecological, and cosmological connectivity's potential to foster transformative action compare (O'Sullivan, 1999; 2005)

> proof. Here the burden of proof is effectively reversed, since in such cases objective and conclusive demonstrations will have to be brought forward to demonstrate that the proposed activity will not cause serious harm to the environment or to those who inhabit it (Francis, 2015, #186).[8]

In the spirit of this understanding of science and technology as properly directed towards supporting socio-ecological flourishing, Francis summarizes his teaching on the technocratic paradigm by arguing that rather than unreflectively accepting elements of an unjust status quo we ought to embrace:

> the urgent need for us to move forward in a bold cultural revolution. Science and technology are not neutral; from the beginning to the end of a process, various intentions and possibilities are in play and can take on distinct shapes. Nobody is suggesting a return to the Stone Age, but we do need to slow down and look at reality in a different way, to appropriate the positive and sustainable progress which has been made, but also to recover the values and the great goals swept away by our unrestrained delusions of grandeur (Francis, 2015, #114).

These criteria, related to coming together of science and technology, are important to the analysis of the challenges and promises of radical life extension presented below in this chapter. More broadly, *Laudato Si'* itself is, in many ways, centered upon grounding integral ecology via multilevel learning and action in order to better care for our common home. For Francis (2015), science can act as a catalyst for such transformative integration. Indeed, that process is facilitated when science is pursued in light of the insights of integral ecology. At a minimum, for secular scientists this means placing their findings within social, political, ecological, and cosmological contexts, which are herein conceived as intimately relational.

For those subscribing to corresponding theo-ethical premises, the pursuit of socio-ecological flourishing that flows from the grounding of integral ecology means that the prospects for the incarnation of social justice and a creative common good are negatively affected by any disruption of human-Earth-divine relationships. Moreover, the full consequences of harms to be addressed in order to adequately participate in caring for our common home can only come into view with the help of science exercised within a framing of

---

[8] For a theo-ecoethical consideration of the precautionary principle, see Deane-Drummond (2008).

deep relationality that is inseparable from integral ecology. As has begun to be demonstrated in this chapter, such illustrations are indicative of how 'integral ecology' as a concept serves to highlight that while science has been a source of the challenges facing the Earth community, it is also essential to an increasingly necessary and urgent human role in responding to intertwined social and ecological crises. Further, it is proposed that a moral project of helping to foster a transformative socio-ecological flourishing provides an appropriate standard of practice for the scientific project at this stage in planetary history.

Can human enhancement be a path whereby science promotes socio-ecological flourishing?

Addressing comparable challenges related to multi-dimensional sustainability, futurist Ray Kurzweil sites a number of developments that, due to their divergent foundations, seemingly turn Pope Francis's views on rapidification on their head. At the inaugural Conference of the Saskatchewan Centre for Religion and Science addressing the topic of "Spiritualities of Human Enhancement and Artificial Intelligence" in December 2017, Kurzweil explained:

> Recently, in Asia they demonstrated the feasibility of putting together a three-story office building in three days using little module they slap together Lego style. These components were printed on a 3D printer. That will be the nature of construction in the 2020s. Most of our physical things will come from 3D printing. This development will also mark the beginning of a vertical agriculture revolution where we can grow high-quality food without chemicals in vertical buildings. This approach takes up one percent of the land required by horizontal agriculture. So, as a result of the deflation rate we will be able to produce high-quality food very inexpensively. Hydroponic plants for foods and vegetables, in-vitro cloning of muscle tissue for meat. People for the Ethical Treatment of Animals (PETA), the animal rights group is not opposed to people eating meat as long as it does not come from animals. The first hamburger to be produced this way has been consumed. It was expensive, a few hundred thousand dollars. But similarly, this will be subject to the deflation rate and ultimately be very inexpensive. The hamburger was very good by the way. Indeed, virtually all the industries are going to be transformed as information technologies, once we get these enabling factors (Kurzweil, 2017).

Read through the relational lens of integral ecology, this series of developments can at first seem very promising in terms of the human footprint within the larger life community. Adapting a common argument of vegan animal rights activists who question how love for animals is compatible with eating them (e.g., Cyrus, 2017), within the vision offered by Kurzweil, we can better assert our love for our more-than-human animal brothers and sisters by not consuming them. Further, we can reduce our chemical and transportation footprint, making more space for other members of the Earth community to flourish.

Due to human health being related to ecosystem health, in and of itself the accompanying reduced use of toxicants may extend human lifespan. However, the tethered technological developments also retain numerous tensions in relation to Francis' teaching on the above-mapped need to humanize technoscience within the reality of deep connectivity pointed to by his articulation of integral ecology. As will come into focus below, this note of caution takes on particular nuances when the energies of technoscience are focused on human enhancement. Of course, even eyeglasses can be read as a mode of ameliorating the bodily condition, and corresponding broader definitions of human enhancement are worth considering in terms of their ethical and social impacts (see, for example, Trothen, 2015). In the broader areas of transhumanism, such considerations led Cudworth and Hobden (2018) to articulate a political project for critical 'terraists' politics to overcome the 'capitalocene'. Significantly for the subject matter of this chapter they, in part, base their political program on a needed "(re)discovery in the joy of living in our shared morality" (Cudworth and Hobden, 2018, 158). The present chapter proceeds with a more modest scope but in a comparable spirit to that displayed by Trothen (2015) and Cudworth and Hobden (2019). More specifically, this chapter will proceed by combining a consideration of the social logic and ethical consequences of human enhancement as it undertakes a brief survey of select recent trends in radical life extension.

In addition to his inventiveness, gripping predictions for the future, and skills with computer programming and technology, Kurzweil is also a well-known proponent of radical life extension. Indeed, his naming of the possibilities in this regard has a particular appeal precisely because it combines insights from each of the areas for which he has gained considerable notoriety (see, for example, Kurzweil, 2017). There are many proposals seeking to increase human lifespan well beyond our current experience of embodiment. For instance, radical life extension's prospects could be improved by nanorobots attacking disease and the effects of aging from within on our bodies. Such applications promise much longer lifespans (Kurzweil, 2007). This may be problematic because prolonging human life increases our abilities to consume. Further, at least initially, it is elites who would benefit from the enabling technologies.[9] However, for those seeking Jewish and Christian scriptural precedents to justify radical life extension, they can be found in the Book of Genesis' (ch. 5) listing of the lifespans of the early patriarchs. In that section of the Pentateuch, several figures, including the original human, Adam, are named as living close to a millennium. These long lifespans frequently attributed to a closeness to the divine that was understood as having been reduced over human history as a result of the compounding reality of human sin in the world (see Hill, 2003; Stump 2017). Today, there is an idea that genetic manipulation combined with medical nanorobots (see Li et al., 2017) could extend life to recover these biblical proportions. In a departure from Pope Francis' (2015) treatment of the relevant concepts, this ambition recalls "the orientation of one of" after the key anglophone pioneers of modernist science, then explicitly mixing theological themes, was an active consideration in Francis Bacon's (1638/2010) notion of overcoming the effects of sin through the progress of science as active in the London-based Royal Society's foundational period.[10]

A related set of issues emerge when radical life extension is considered in terms of possible developments in cyborg technologies. For example, some cyborg enthusiasts seek to manifest a promise of a certain type of connection to transcendence that retains personal individuality and is correlated with the doctrines of many Western religious traditions. In an important departure, those affiliating with this strand of radical life extension seek to actualize even longer-than-biblical lifespans by overcoming biological limitations through the anthropogenic 'creation' of trans-and post-humans (see Vicini and Brazal, 2015). This effectively transcendental aspiration brings a particular focus to bear upon the ultimate goals for some transhumanists: a disembodied, de facto immortality that somewhat ironically clings to this present existence, which this present chapter demonstrates is to this point in our history shaped by our biological limits and possibilities.

Yet, this is a very different kind of clinging, one that, if successful, will add a new dimension to Earthy existence as it works to extend personal lifespans beyond the one-thousand-year-old threshold.[11] As a result, it aspires to overtake even the ultimate biblically-recorded achievement of Methuselah who is named in the Book of Genesis (5:27) as having lived a remarkable 969

---

[9] On the issues of access and equity in play here compare Deane-Drummond (2006/2010).

[10] On the relationships between sin and Francis Bacon's ambitions for science, compare Deane-Drummond (2006/2010) in conversation with Greenburg (2017).

[11] See Maher and Mercer (2009) on some of the multi-religious implications in play when striving to manifest radically longer human lifespans.

years. This sought-after achievement has correspondingly been 'christened' the 'Methuselarity', in a nod to the biblical example, by Paul Hynek, and brought into radical life extension conversations by his friend the biogerontologist Aubrey de Grey (2009), whose thought Hynek had been seeking to encapsulate when he came up with the term. Extolling the merits of regenerative medicine to overcome aging, de Grey (2009) defines the Methuselarity as "the point in our progress against aging at which our rational expectation of the age to which we can expect to live without age-related physiological and cognitive decline goes from the low three digits to infinite. And [writing in the first person] my use here of the word 'point' is almost accurate: this transition will, in my view, take no longer than a few years" (198). This framing follows upon a comparable but less biologically-oriented and more technoscience-influenced use of the term, 'the singularity'.

Notably, Ray Kurzweil (2005) has employed that term to indicate a point where the exponential growth of artificial intelligence melds with superhuman intelligence, effectively rendering a fused form of immortality possible as distinctions between humans and machines melt away. Earlier, Kurzweil (1999) connected the anticipated future consciousness of artificial intelligence with a spirituality, as part of his commentary in this regard he posited that computers would meditate and attend religious institutions to explore that spirituality once they felt they had consciousness. Given such reflections and the general ambitions in play in the quest for radical life extension, it is not surprising to see the emergence of religious transhumanists, who develop theologies and spiritualties that stand in contrast to the principles of integral ecology as articulated by Pope Francis. Yet, at the same time these theologies can challenge the assumptions of famous thinkers like the cyborg theorist Donna Haraway (see 1991/2010). This challenge has been affected, for example, by articulating a 'cyborg theology' drawing on Jewish and Christian sources, a prospect which appeared incompatible with cyborg culture to Haraway (see Midson, 2018).

Focussing on human immortality on a more biological level, which initially appears to be a better approach from a perspective influenced by integral ecology, and based upon a rubric of proper maintenance as rejuvenation (see de Grey and Rae, 2008), de Grey (2000) also promotes an idea of 'longevity escape velocity'. That term is indicative of the point after which medical advances will be able to repair the damage caused to cells over time in order to prevent biological aging. From a religious studies perspective, the religious and post-religious elements of such nomenclature as applied to radical life extension are obvious. Indeed, in that light, the fuller meanings of this nomenclature make little sense without a particular other-worldly oriented religious background to the cultures from where they are articulated. As such, for example, the ambitions of some radical life extension advocates take on shades, and even appropriate the language of 'eternal life'. Yet, this appropriation results in articulations of eternal life that are distinctively anthropogenic. From a theo-ecoethical point of view, the connections between (1) what might thus be labeled 'anthropogenic eternal life' and (2) what David Kinsley (1996) has termed "the mastery hypothesis" are telling here (116).

In short, the mastery hypothesis represents an almost pure anthropocentrism in human-Earth relationships, which is manifested in the idea that because humans are made in 'the image and likeness of God' (compare Genesis 1:26-28) they are to be considered above all other creatures. In this view, creation and all 'its bounty' is for humankind to use as it chooses. In a revealing (though condescending) analogy, Janet Martin Soskice (1995) labels such sentiments as 'divine hamster cage anthropocentrism'. Within such an ethical framework, "God is the hamster owner and we are the hamsters. God creates the world as a kind of vivarium for human beings. The rest of the created order is our lettuce leaves and clean sawdust, completely at our disposal" (Soskice, 1995, 64)

Although ecologically-oriented biblical scholars (for example, Sharkey, 2004) and Pope Francis (2015) offer alternative readings, those who support the mastery hypothesis may look on the granting of "dominion" over the natural world that occurs in Genesis 1: 28 as a divine ordinance providing justification for subduing and exploiting the Earth (Carmody, 1983, 6). While based on what may be characterized a mishearing,[12] this understanding nonetheless translated into a situation where, for the most part, not only Christians but most Westerners could regrettably feel "quite self-righteous in their quest to tame, civilize and otherwise dominate nature even if that meant destroying large parts of it" (Kinsley, 1996, 116). In theo-ecoethical terms, such a hierarchical image serves to remind us of the limits on freedom and grace that accompany an overly utilitarian approach to nature as spaces for action and relationship building are unduly limited. Rather problematically, ambitions for anthropogenic eternal life may serve to further remove the divine from the picture. They can do so by extending the purview of human domination over nature via myopic applications of progress and technique, thus wholly

[12] Pope Francis (2015) is adamant in *Laudato Si'* that Genesis 1:28 should not be read as granting a licence for exploitation of more-than-human members of the created community. The scientifically literate Cardinal Cahal B. Daly (2004) argues along compatible lines that the relationship named in Gen. 1:26-28 is one of custodianship that requires people to care for our "common and fragile home" understood to exist as part of God's good creation (109, 222).

removing the promises and perils of cosmic justice from the moral equation in human-Earth relations. As a result, radical life extension undertaken along such lines becomes just another mode of mastery seeking to erase natural limits.

Yet, given radically extended lifespans, one might expect a certain slowing down in terms of the pace of life, so that ecological footprints and the urge to master nature in haste, at least in real-time, are reduced. However, the vision for getting to that point relies on the exponential growth in technoscience that makes life busier for those caught within its wide-reaching net and enables many aspects of the exploitative practices responsible for the current 'socio-ecological crisis'.[13] Once again in contrast to Pope Francis' articulation of a critique of rapidification, Kurzweil is also well known for predicting,[14] based upon continuous exponential growth in technology that by 2045 we will be able to download our individual and unique consciousnesses into entirely non-biological carriers. Seeking to realize that goal is Russian billionaire Dmitry Itskov's *2045 Initiative*. Founded in 2011, this initiative has targeted those who appear on the *Forbes* List to provide the financial battery for its *Project Avatar*. The *2045 Initiative* also includes an interfaith dialogue dimension that piggybacks on Kurzweil's predictions, while employing transcendental terms capturing the intersections prevalent when hyper technology meets spiritualities of human enhancements and transhumanism in search of a radically extended personal life. In this regard, note the way that 'immortality' is treated in the following paragraph, as Itskov's organization employs scientific framing and language that strikes a chord with Abrahamic traditions but does so by bypassing a role for divine action:

> The main science mega-project of the *2045 Initiative* aims to create technologies enabling the transfer of an individual's personality to a more advanced non-biological carrier, and extending life, including to the point of immortality. We devote particular attention to enabling the fullest possible dialogue between the world's major spiritual traditions, science and society (2045 Initiative Staff, 2012).

Some have suggested that a comparable transfer of consciousness approach to such enabling of anthropogenic eternal life will be a technique for allowing purposely disembodied humans to survive the end of a vital planet Earth when our sun becomes a red giant. They have also upheld this form of transfer

---

[13] For inspiration for the use of the term 'socio-ecological crisis' in this chapter see Eaton (2013).

[14] For example of the way this aspect of Kurzweil's work has been received see Lewis (2013).

of consciousness as even permitting individual people to navigate the heat death of the universe. For example, in Freeman Dyson's (1979; 1988/2004) discussion-starting formulation, a combination artificial and natural intelligence would evolve to allow consciousness transferred to hardware in dust clouds floating around the cosmos to survive for trillions of years at ultra-low temperatures. Dyson introduced this possibility of transfer of consciousness in his Gifford Lectures on religion and science as follows:

> My argument will be based on a fundamental assumption concerning the nature of life, that life resides in the nature of organization rather than in substance. I am assuming that consciousness is inherent in the way the molecules are organized in my head not in the molecules themselves. If this assumption is true, that life is organization rather than substance, then it makes sense to imagine life detached from flesh and blood as existing in networks of superconducting circuitry or in interstellar dust clouds (Dyson, 1988/2004, 107).

Hardware enabling this type of existence would use little energy and, thus, from a theo-ecoethical perspective, would manifest the ultimately minimal ecological footprint for a rational being. Nonetheless, in a tension following upon his assumptions about the nature of life as ultimately unsustainable, should we arrive at such a point, despite Dyson's somewhat misleading invocations of (network) connectivity and embodiment detached from flesh and blood in the above quotation, we will have lost something that matters from an ecospiritual perspective: our presently definitive embodied finitude. As such, one could conclude integral ecology will be irrelevant due to our new location in spacetime. However, the insights underlying integral ecology also imply it might be that we will have bypassed something essential to our human identity. Moreover, this conclusion does not rest solely on the problematic transhumanist ambition that our time in the material universe, supposedly in this very existence, will continue indefinitely. Rather, integral ecology highlights the possibility that there may be something essential about connectivity via the very flesh and blood that the likes of Kurzweil and Dyson leave behind when they seek to put forward notions that, for the sake of longevity, we can, and perhaps even should, leave our bodies behind. Yet, while that disembodiment may hold promise for sustainability in terms of low energy use to maintain particular formulations of personal conscience, it also means that we lose the chance to contribute to social-ecological flourishing in a tangible and substantive manner as fleshy participants.

Ecological conversion and embodiment matters

Daniel Castillo (2016) has argued persuasively that, as articulated by Pope Francis, "the concept of integral ecology should be construed as a liberationist concept, one that calls for a paradigm shift away from the structural and ideological dimensions of the contemporary 'global system'" (354). Castillo (2016) continues "simply put, in calling for an integral ecology, Francis is calling for the radical conversion of the entire global system" (371). This chapter had named some of the problematic intersections between elitism and transhumanism. It has also served to underline that this liberation is not a disembodied otherworldly one but, rather, is grounded in the universe-embedded and relational context of the Earth community. Even prior to the release of *Laudato Si',* Pope Francis gave several nods to such embedded ecological wisdom in his first apostolic exhortation, *Evangelii Gaudium.* The theological anthropology he presents notes humanity's embodied nature. Francis (2013) further teaches that our embodiment connects us to the rest of the natural world on an immediate and sensory level: "Thanks to our bodies, God has joined us so closely to the world around us that we can feel the desertification of the soil almost as a physical ailment, and the extinction of a species as a painful disfigurement" (Francis, 2013, #215). Giving further support to a green reading of the importance of embodiment for integral ecology, Francis has positioned science and technology as properly directed towards supporting socio-ecological flourishing. Here, drawing on his most well-known contemporary predecessor, the Pope complements his teaching on the technocratic paradigm by arguing that we ought to participate in an ecological conversion. Invoking a poignant image of humanity's proper role in the universe, Pope John Paul II (2001) taught that in recent decades "ecological conversion....has made humanity more sensitive to the catastrophe to which it has been heading." He continued that the human "is no longer the Creator's 'steward', but an autonomous despot, who is finally beginning to understand that he must stop at the edge of the abyss" (John Paul II, 2001). Mediated by Francis' articulation of integral ecology, this is an ethical turning to see both our essential location as Earth creatures and the corresponding duty, that is ultimately in our interest, to support the flourishing of people living in poverty and the health of Mother Earth.

This concept of ecological conversion is further situated, in accord with his broader vision, when Francis argues that realizing the common good under conditions of ecological crisis means that insular approaches in their manifestations within religions, disciplines, or social movements are no longer tenable. The Pope expands on this point with specific reference to the sciences:

> The majority of people living on our planet profess to be believers. This should spur religions to dialogue among themselves for the sake of protecting nature, defending the poor, and building networks of respect and fraternity. Dialogue among the various sciences is likewise needed, since each can tend to become enclosed in its own language, while specialization leads to a certain isolation and the absolutization of its own field of knowledge. This prevents us from confronting environmental problems effectively. An open and respectful dialogue is also needed between the various ecological movements, among which ideological conflicts are not infrequently encountered. The gravity of the ecological crisis demands that we all look to the common good, embarking on a path of dialogue which demands patience, self-discipline and generosity, always keeping in mind that [as Francis (2013) teaches in *Evangelii Gaudium*] "realities are greater than ideas" (2015, #201).

Although, it should be acknowledged that this sentiment is somewhat tempered in terms of Christian dialogue with Asian religions and natural scientists who may take offense to the 'us vs. them' style framings and emic faith-inspired metaphysical sentiments that Francis (2015) at times expresses in *Laudato Si'*. For example, such tempering is given expression when he teaches, "how wonderful is the certainty that each human life is not adrift in the midst of hopeless chaos, in a world ruled by pure chance or endlessly recurring cycles!" (#65).

As one significant correlate in this regard, *Laudato Si'* develops the papal contribution to ecospirituality in a more integral and, when viewed through a theo-ecoethical lens, markedly innovative direction.[15] Francis (2015) does so by arguing that a reality of connectedness as accessed through ecological conversion provides creative energy of the sort that is necessary to effectively respond to contemporary challenges facing the Earth community: "As believers, we do not look at the world from without but from within, conscious of the bonds with which the Father has linked us to all beings. By developing our individual, God-given capacities, an ecological conversion can inspire us to greater creativity and enthusiasm in resolving the world's problems and in offering ourselves to God" (Francis, 2015, #202). This is certainly the type of conversion that opens those who participate in it to the treasures of an embodied existence lived in an intentional way within the Earth community. Correspondingly, *Laudato Si'* itself is in many ways centered upon grounding integral ecology via multilevel learning and action in order to better care for our

---

[15] On papal ecospirituality prior to Pope Francis' reign see Hrynkow and O'Hara (2011).

common home. Indeed, it is this entangled ethos of care for all creation that is the ultimate ethical outcome of any substantive ecological conversion. For Francis, science can act as a catalyst for such transformative integration. Such transformation is facilitated when science is pursued in light of the insights of integral ecology. At a minimum, for secular scientists, this means placing their findings back within social, political, ecological, and cosmological contexts, which are thus conceived as intimately relational.

Here we come to a significant note of caution when technological progress seeks to take elite humans out of their embodiment. Less any ambiguity remain, it should be emphasized that, for the Pope, our embodiment ties us to the Earth community in a deeply relational manner that intimately and irreversibly joins us to the rest of creation in a manner that ought to temper domineering impulses:

> It is enough to recognize that our body itself establishes us in a direct relationship with the environment and with other living beings. The acceptance of our bodies as God's gift is vital for welcoming and accepting the entire world as a gift from the Father and our common home, whereas thinking that we enjoy absolute power over our own bodies turns, often subtly, into thinking that we enjoy absolute power over creation. Learning to accept our body, to care for it and to respect its fullest meaning, is an essential element of any genuine human ecology. Also, valuing one's own body in its femininity or masculinity is necessary if I am going to be able to recognize myself in an encounter with someone who is different. In this way we can joyfully accept the specific gifts of another man or woman, the work of God the Creator, and find mutual enrichment (Francis, 2015, #155).

Notwithstanding the rather dichotomist and essentialist treatment of gender in this passage, it is a result of that deep, embodied connectivity so important to integral ecology and flowing from Pope Francis' theological anthropology, authentic human development is firmly placed within a moral perspective that is at once respectful of people and the rest of creation. For those accepting this line of argument within a spiritual worldview, the pursuit of socio-ecological flourishing that flows from the grounding of integral ecology means that human-Earth-divine relationships have both vertical and horizontal dimensions that correlate with moral duties to duties to God and neighbor. In terms of humanity's duty to contribute to socio-ecological flourishing, we owe many debts to the Earth community. After taking so much we can at a minimum, from a view inspired by integral ecology, give our bodies back to the Earth so that they can become actual food for worms and microbes (at least, theologically speaking, until the resurrection of the body takes place).

Otherwise, in failing to attend to the multiple debts humans we owe to the Earth community that demand much more than the mere return of our bodies to dust from whence we came, we risk continuing the trend of using more than our fair share of resources. The cogent goal set by Francis (2013, 2015) here can be framed as fostering a substantive, embodied socio-ecological flourishing that is marked by human equality. As such, it is especially morally problematic from a perspective influenced by integral ecology if only elites gain access to radical life extension. In this light, the question that Francis (2015) asks about what kind of world we are leaving for our children becomes doubly cogent when aspirational elements of radical life extension would leave us as consumers. Moreover it is elites, most of whom are from the West and are already participating in cultures of overconsumption, who are seemingly positioned as the first cohort of humans to benefit from such technologies, extending a high-intensity ecological footprint over much larger swaths of spacetime. From a perspective inspired by Pope Francis' articulation of integral ecology, it must therefore be emphasized that it is the ultimate expression of some of the most troubling aspects of the technocratic paradigm to 'save' only the elites from death.

As is evidenced in this chapter, such illustrations are indicative of how 'integral ecology' as a concept serves to highlight that science has been a source of the challenges facing the Earth community. Its normative elements illustrate that science is also essential to an increasingly necessary and urgent human role in responding to intertwined social and ecological crises via helping to foster a transformative socio-ecological flourishing. In this light, devoting time and resources to radical life extension may be viewed as morally problematic on each of the aforementioned levels of Earth-human-divine relationships, misdirecting resources and time that may be better spent on cultivating the diverse ways of being just in the world that are so crucial to Pope Francis' articulation of integral ecology.

Conclusion: what we leave behind, socio-cosmic democracy, and the precautionary principle

"What we leave behind" is a significant framing within the field of faith-inspired ecological ethics as it relates to the scientific project's intersections with our duties to future generations within the context of human and larger life communities (see, for example, Jensen and McBay, 2009). This insight rings as true precisely because of scientists' location within those communities. Thus, without much turbulence that framing can be applied to the ethical problem of what we leave behind if humans, understood as embodied creatures, seek to transcend that location, which places us as Earth creatures. The similarities between transhumanists and more Gnostic or platonic versions of Christianity that problematically downplay the importance of the body are eerily obvious (see Pugh, 2017). Here, if visions like Dyson's (1979; 1988/2004) aforementioned proposals are realized it must be conceded that we will make quantum advances in terms of sustainability as it relates to energy use. Yet, from the perspective of Pope Francis' articulation of integral ecology accompanied by an anthropology of creaturely embodiment that together views us as essentially situated within a created common home, transhumanist ambitions as they pertain to radical life extension also represent the potential for a problematic loss.

This certainly does not discount Kurzweil's (2017) general point that people, presumably on the level of consciousness, could adapt to much longer lifespans. He makes an analogy here about how they have managed to adapt to more commonly living into their 80s and 90s in affluent societies, despite that fact that when within (what is often) living memory their ancestors could only expect to live into their 30s and 40s (Kurzweil 2017). Still if victory over death is achieved on the biblical scale proposed in some of the formulations surveyed above, a perspective influenced by Pope Francis' articulation of integral ecology holds out the possibility that such an accomplishment could represent a Pyrrhic victory. This remains true despite some of the real advantages for sustainability in terms of personal energy use that may accrue. Applied to such analysis, integral ecology provides an analytical concept to question some telling social and ecological senses in which the cost of a this-worldly personal victory over death may simply be too high. Here, Francis' teachings beg us to consider the ethical consequences of the shifts in relationality that would be inherent within a situation where everything is connected if such a reality comes into being. In short, personal radical life extension will have systematic impacts on a relational level that should be not be summarily dismissed or ignored as inconsequential.

In this regard, it is instructive to consider how some ecological ethicists employ the term 'more-than-human' to de-construct what are often presently normative anthropocentric hierarchies that place *Homo sapiens* above other beings. As footnoted above, this term is also meant to indicate that respect is due to the members of the Earth community who are not human persons (see Abram, 1996/1997). Yet, many transhumanists, as a consequence of their ambitions, aspire to be more-than-human in accord with a plainer meaning of that term, one emptied of the Earth community context. This is a significant difference because, from an integral ecology perspective, the larger life community has shaped personhood and otherwise set boundaries on *Homo sapiens*' existential experience of this-worldly existence to this point in our species' history. Minimally, this will change the Earth-embedded rules and limits that have hitherto defined human life. Like other developments, including nuclear technology (to cite a science-infused example), people's inventiveness in this regard will be accompanied by tensions and promises, with the former magnified when that technology is not humanized in the integral sense named by Pope Francis (2015) and as unfolded above. Indeed, like the nuclear bomb changed high-technology warfare, extending life to and even beyond the sort of biblical proportions that attracted Francis Bacon will quite simply shift the rules of the proverbial game of human life as we have known them. While, even for many beyond transhumanist circles, such change may hold appeal, perhaps the best response to these promises and tensions from an ecospiritual point of view is a 'socio-cosmic' democratic one.

Referring to the fusion of social and cosmological consciousness and action, as with the concept of integral ecology, 'socio-cosmic democracy' was introduced into Catholic social thought by Leonardo Boff (see 2011). To synthesize that concept while employing some of the key terms unfolded in this chapter, within a socio-cosmic democratic approach the interests of all are seen as connected within a rubric of integral ecology. This remains true for a wealthy elite minority, the poorest human communities or the most endangered species. Herein, the interest and duties of all are tethered to a duty to support socio-ecological flourishing in such a manner that favors marginalized communities so that they can participate in that flourishing. Reflectively working to incarnate the imperatives for this liberationist and widely emancipatory outcome, a socio-cosmic democratic approach allows for a fuller consideration of whether the ambitions of radical life extension advocates represent what is truly going forward (see Lonergan 1972/2017) at this juncture in planetary history. This reflective ethical program is important from a viewpoint influenced by integral ecology because our present reality is marked by the aforementioned compounding of the socio-ecological crisis that is itself underplayed or sidestepped in the thought of many transhumanists. As such and from a perspective informed by Pope Francis' articulation of integral ecology, radical life extension advocates, like all scientists and technologists, need to be careful to integrate the insight that the present socio-ecological crisis has relational consequences that are felt by people in their bodies. As such, if embodiment matters in anything approaching the ecospiritual senses unfolded above, then a theo-ecoethical anthropology underpinning socio-cosmic democracy requires pause for thought and regulation of the modernist scientific project in so much as it forgets the relational context of the Earth and universe communities. Further, to the degree that the ambitions of radical life extension advocates and technologists amplify rapidification and display symptoms of a comparable amnesia, they also require reflection and regulation.

These conclusions follow precisely because consideration of the context in an integral regard has social and environmental consequences that bring Francis' (2015) above-cited teaching on the precautionary principle into play in support of the imperatives of socio-cosmic democracy. Applications of the precautionary principle along these integral lines mean accepting the premise that due caution is necessary so as to avoid the consequences of the unchecked ambition of radical life extension advocates flowing over into game-changing shifts for human embodiment as we have known it. This application of the precautionary principle is particularly important because it is highly probable that at least some of the concomitant outcomes will be undemocratic in a socio-cosmic sense. Yet, caution is not enough in relation to the scientific project writ large from a perspective informed by Pope Francis articulation of integral ecology; in terms of this chapter there is also the positive moral duty to contribute to our common home becoming a fuller place of socio-ecological flourishing.

In this sense, integral ecology points back to an imperative for all people engaged in biomedical and technoscience work to account for their projects in terms of their contributions to socio-ecological flourishing. This accounting may be within the realm of the possible for radical life extension advocates. However, if the perspective on integral ecology unfolded in this chapter is taken seriously, there are certainly enough tensions active in the realm of radical life extension that make that exercise of the scientific project a cause for deep ethical concern. As a result and in accord with the precautionary principle, before the human community gives their assent to technoscientific and biomedical approaches to radical life extension, researchers must conclusively demonstrate that there is a high probability that the outcomes of their approach will actively benefit an integral common good. The problem, of course, is that such consent is rarely, if ever, sought and, thus, technoscience proceeding in line with segmented and myopic interests remains un-humanized in the integral sense discussed above. In contrast, socio-cosmic democracy holds up the necessity of deep participation to address this democratic deficit. Integral ecology as articulated by Pope Francis (2015) invites everyone to join in that effort, so as to enter into conversation across lines of division via the realization of common origins and a sense of belonging to the Earth community. In the terms brought into play in this chapter, the resulting hope is that a vital future, which can truly be shared amongst all earthlings, will be co-constructed in socio-cosmically democratic manner.

To close, it is important to emphasize that this invitation unambiguously includes all radical life extension advocates. That invitation is pertinent even if they have to ultimately leave behind some of their ambitious, through a process of ecological conversion, in order to contribute to socio-ecological flourishing. In light of such a goal flowing from Pope Francis's articulation of integral ecology as mapped in this chapter, who and what we may leave behind in term of our location as presently embodied and relational beings matters in several profound senses.

References

2045 Initiative Staff., 2012. About us. *2045 Strategic social initiative.* 17 July. Available at: < http://2045.com/about/> [Accessed 1 August 2018].

Abram, D., 1996/1997. *The spell of the sensuous: Perception and language in a more-than-human world.* New York: Vintage.

Bacon, F., 1638/2010. *The history of life and death.* Whitefish, MT: Kessinger Publishing.

Benzoni, F. J., 2007. *Ecological ethics and the human soul: Aquinas, Whitehead, and the metaphysics of nature.* Notre Dame, IN: University of Notre Dame Press.

Berry, T., 2002. Interview with Thomas Berry. In: D. Jensen. 2004. *Listening to the land: Conversations about nature, culture and eros.* White River Junction, VT: Chelsea Green Publishing Company. pp. 35-44.

Berry, T., 2006. *Evening thoughts: Reflecting on Earth as sacred community.* Mary Evelyn Tucker, ed. San Francisco, CA: Sierra Club Books.

Boff, L., 1995. *Ecology and liberation: A new paradigm.* Translated by J. Cumming. Maryknoll, NY: Orbis.

Boff, L., 1997. *Cry of the Earth, cry of the poor.* Translated by P. Berryman. Maryknoll, NY: Orbis.

Boff. L., 2011. *Virtues for another possible world.* Translated by A. Guilherme. Eugene, OR: Cascade Books.

Carmody, J., 1983. *Religion and ecology: Toward a new Christian theology of nature.* New York: Paulist Press.

Castillo, D.P., 2016. Integral ecology as a liberationist concept. *Theological Studies,* 77 (2), pp. 353-376.

Chipman, S., 2017. Quoted in Crusader Staff. 2018. Steve '68 and Ruth Chipman honoured with 2017 Ignatian challenge award. *The Crusader,* Winter issue, pp. 8-9 [online]. Available at: <https://issuu.com/stpaulshighschool/docs/the_crusader_-_winter_2018_-_web> [Accessed 22 July 2018].

Cudworth, E. and S. Hobden., 2018. *The emancipatory project of transhumanism.* London: Routledge.

Cyrus, M., 2017. If you choose to eat meat .... you love PETS not ANIMALS..... *#loveanimalsdonteatthem #vegan RG.* 8 July. Available at: <https://twitter.com/MileyCyrus/status/883730739828711424> [Accessed 23 July 2018].

Daly, C.B., 2004. *The minding of planet earth.* Dublin: Veritas.

de Grey, A.D.N.J., 2009. The singularity and the Methuselarity: similarities and differences. *Studies in Health Technology and Informatics,* 149, pp. 195–202.

de Grey, A.D.N.J. and M. Rae,, 2008. *Ending aging: The rejuvenation breakthroughs that could reverse human aging in our lifetime.* New York: St. Martins Griffin

Deane-Drummond, C., 2006/2010. Future perfect? God, the transhuman future and the quest for immortality. In C. Deane-Drummond and P. M. Scott, eds. *Future perfect? God, medicine and human identity.* London: T&T Clark, pp. 168-182.

Deane-Drummond, C., 2008. *Eco-theology.* Winona, MN: Saint Mary's Press.

Dicks, H., 2016. The philosophy of biomimicry. *Philosophy & Technology,* 29(3), pp. 223-243.

Drummond, K., 2012. Russian mogul to "Forbes" billionaires: Limitless lifespans can be yours. *Forbes* [online] last updated at 2:18 on 18 June 2012. Available at: <https://www.forbes.com/sites/katiedrummond/2012/07/19/dmitry-itskov-avatar/#6d14980157ad> [Accessed 3August 2018].

Dyson, F.J., 1979. Time without end: Physics and biology in an open universe. *Reviews of Modern Physics,* 51(3), pp. 447–460.

Dyson, F.J., 1988/2004. *Infinite in all directions: Gifford lectures given at Aberdeen, Scotland April-November 1985.* New York: First Perennial.

Eaton, H., 2013. Forces of nature: Aesthetics and ethics. In: S. Bergmann, I. Blindow and K. Ott, eds. *Aesth/ethics in environmental change: Hiking through the arts, ecology, religion and ethics of the environment.* Berlin: LIT Verlag. pp. 109-126.

Francis., 2013. Apostolic exhortation *Evangelii Gaudium* of the holy father Francis to the bishops, clergy, consecrated persons and the lay faithful on the proclamation of the gospel in today's world [online]. © Libreria Editrice Vaticana. Available at: <http://w2.vatican.va/content/francesco/en/apost_exhortations/documents/papa-francesco_esortazione-ap_20131124_evangelii-gaudium.html> [Accessed 22 July 2018].

Francis., 2015. Encyclical letter *Laudato Si'* of the holy father Francis on care for our common home [online]. © Libreria Editrice Vaticana. Available at: <http://w2.vatican.va/content/francesco/en/encyclicals/documents/papa-francesco_20150524_enciclica-laudato-si.html> [Accessed 22 July 2018].

Gaillardetz, R. R., 1997. *Teaching with authority: A theology of the magisterium in the church.* Collegeville, MN: Liturgical Press.

Gaillardetz, R. R., 2003. *By what authority?: A primer on scripture, the magisterium, and the sense of the faithful.* Collegeville, MN: Liturgical Press.

Greenburg, A.L., 2017. Francis Bacon's new science of life-prolongation. *Synapsis: A Health Humanities Journal* [online]. Available at: <https://medicalhealthhumanities.com/2017/11/08/francis-bacons-new-science-of-life-prolongation/> [Accessed July 30, 2018].

Haraway, D. J., 1991/2010. *Symions, cyborgs, and women: The reinvention of nature.* New York: Taylor & Francis.

Hill, C.A., 2003. Making sense of the numbers of genesis. *Perspectives on Science and Christian Faith,* 55(4), pp. 239-251.

Hogue, M.S., 2010. *The tangled bank: Toward an ecotheological ethics of responsible participation.* Cambridge: James Clarke and Co.

Hrynkow, C., 2014. Transforming cultures, growing substantive peace: Pope Francis, St. Francis, and a green peace. *Peace Studies Journal*, 7(4), pp. 25-38.

Hrynkow, C., 2016. *Laudato Si'*, transformative learning, and the healing of human-Earth-divine relationships. *The Ecumenist: A journal of theology, culture, and Society*, 53(4), pp. 10-15.

Hrynkow, C., 2017. Greening God? Christian ecotheology, environmental justice, and docio-ecological flourishing. *Environmental Justice*, 10(3), pp. 81-87.

Hrynkow, C., 2019. "A true ecological approach always becomes a social approach": A Green theo-ecoethical lens, Pope Francis' teaching, and integral social justice. *Heythrop Journal*, 60(4), pp. 585-599.

Hrynkow, C. and D.P. O'Hara., 2011. The Vatican and ecospirituality: Tensions, promises. and possibilities for fostering an emerging green Catholic spirituality. *Ecozon@: European Journal of Literature, Culture and Environment*, 2(2), pp. 177-197.

Jensen, D. and A. McBay., 2009. *What we leave behind.* New York: Seven Stories Press.

John Paul II., 2001. General audience [online]. 17 January. © Libreria Editrice Vaticana. Available at: <https://w2.vatican.va/content/john-paul-ii/en/audiences/2001/documents/hf_jp-ii_aud_20010117.html> [Accessed 30 July 2018].

Kinsley, D., 1996. "Christianity as ecologically responsible." In: R.S. Gottlieb, ed. *This sacred Earth: Religion, nature, environment.* New York: Routledge, pp. 116-124.

Krieger, N., 2015. The real ecological fallacy: Epidemiology and global climate change. *Journal Epidemiological Community Health*, 69(8), pp. 803-804.

Kurzweil, R., 2007. Opinion, Science. Bring on the nanobots, and we will live long and prosper: We are nearing a tipping point in life extension, thanks to technologies that enhance our health and maximise resources. *The Guardian* [online] last updated at 0:17 GMT on 22 November 2007. Available at: <https://www.theguardian.com/commentisfree/2007/nov/22/comment.comment> [Accessed 30 July 2018].

Kurzweil, R., 1999. *The age of spiritual machines: When computers exceed human intelligence.* New York: Penguin.

Kurzweil, R., 2005. *The singularity is near: When humans transcend biology.* New York: Viking.

Kurzweil, R., 2017. Breaking the Shackles of our Genetic Legacy. Keynote address at *Spiritualities of human enhancement and artificial intelligence* conference. Vancouver, 1 December.

Lewis, T., 2013. The singularity is near: Mind uploading by 2045? *Science Live* [online] last updated at 5:29 ET on 17 June 2013. Available at: <https://www.livescience.com/37499-immortality-by-2045-conference.html> [Accessed 1 August 2018].

Li, J. et al., 2017. Micro/nanorobots for biomedicine: Delivery, surgery, sensing, and detoxification. *Science Robotics*, 2(4), p. 6431.

Lonergan, B.F., 1957/1992. *Insight: A study of human understanding*. F.E. Crowe and R.M. Doran, eds. Volume 3 of the Collected Works of Bernard Lonergan. Toronto: University of Toronto Press.

Lonergan, B.F., 1972/2017. *Method in theology*. R. M. Doran and J. D. Dadosky, eds. 2nd Edition. Volume 14 of the Collected Works of Bernard Lonergan. Toronto: University of Toronto Press.

Maher, D.F. and C. Mercer., 2009. Introduction: Living for 1,000 years—or longer. In: D.F. Maher and C. Mercer, eds. *Religion and the implications of radical life extension*. New York: Palgrave Macmillan, pp. 1-11.

Midson, S.A., 2018. *Cyborg theology: Humans, technology and god*. London: I.B. Tauris & Co.

O'Sullivan, E., 1999. *Transformative learning: Educational vision for the 21st Century*. Toronto, ON: University of Toronto Press.

O'Sullivan, E., 2005. Emancipatory hope: Transformative learning and the strange attractors. In: J. P. Miller, S. Karsten, D. Denton, D. Orr, and I. C. Kates, eds. *Holistic learning and spirituality in education: Breaking new ground*. Albany, NY: State University of New York Press, pp. 69-78.

Piantadosi, S., D.P. Byar, and S.B. Green., 1988. The ecological fallacy. *American Journal of Epidemiology*, 127(5), pp. 893-904.

Pugh, J.C., 2017. The disappearing human: Gnostic dreams in a transhumanist world. *Religions* 8 (5), pp. 51-60.

Knopp, K. et al., 2017. Within- and between-family associations of marital functioning and child well-being. *Journal of Marriage and Family*, 79(2), pp. 451–461.

Rasmussen, L. L., 1996. *Earth community, earth ethics*. Maryknoll, NY: Orbis.

Roszak, T., 1986/1994. *The cult of information: A neo-Luddite treatise on high-tech, artificial intelligence, and the true art of thinking*. Berkeley, CA: University of California Press.

Sharkey, S. A., 2004. *Earth, our home: Biblical witness in the Hebrew scriptures*. Boerne, TX: Sor Juana Press.

Stump, J., 2017. Long life spans in Genesis: Literal or symbolic? *Faith and science seeking understanding* [blog] 5 October. Available at: <https://biologos.org/blogs/jim-stump-faith-and-science-seeking-understanding/long-life-spans-in-genesis> [Accessed 30 July 2018].

Soskice, J. M., 1995. Creation and relation. In: R. Gill, ed. *Readings in modern theology: Britain and America*. Nashville: Abingdon Press, pp. 58-66.

Trothen, T., 2015. *Winnipeg the race? Religion, hope, and reshaping the sport enhancement debate*. Macon, GA: Mercer University Press.

Vicini, A. and A.M. Brazal., 2015. Longing for transcendence: Cyborgs and trans-and posthumans. *Theological Studies*, *76*(1), pp. 148-165.

Wills, T.A. and S.S. Soneji., 2018. Individual-level and ecological studies. *Journal of Adolescent Health*, 62(5), pp. 507-508.

Weber, A., 2016. *The biology of wonder: Aliveness, feeling, and the metamorphosis of science*. New Society Publishers.

Worldviews and artificial intelligence

10.
Possible consequences of AI and transhumanism: Health concerns surrounding unemployment, second class citizenship, and religious engagement

Braden Molhoek
Center for Theology and the Natural Sciences at the Graduate Theological Union

Abstract

This chapter explores the consequences of artificial intelligence (AI) and transhumanism on society. Beginning with how AI and automation contribute to human unemployment, attention is given to Jahoda's notion of the deprivation of the latent functions of employment and how this deprivation can affect health. The impact on AI is also examined, expanding Walzer's understanding of hard work and membership in society to include nonhuman agents. The argument also builds on Herzfeld's argument about humans creating AI in their image and how that relates to the image of God as well as how Jahoda's latent functions might affect AI. The focus then shifts to transhumanism, analyzing the same issues of second class citizenship, unemployment, and the image of God. The chapter concludes with a discussion of religious engagement in light of AI and transhumanism. Religious engagement could help alleviate the deprivation of Jahoda's latent functions and also help avoid viewing others as second class citizens.

Keywords: Transhumanism, Artificial Intelligence, Unemployment, Jahoda's Latent Functions, Second-class citizenship, Image of God

Introduction

The emerging technologies of Artificial Intelligence (AI) and genetic editing that could bring about the transhumanist vision have consequences for society views humans and nonhumans alike. AI and the automation of jobs will lead to increased unemployment for people, leading to real mental health consequences as well as concerns about self-esteem and identity. Looking at existing research on the effects of unemployment, attention is given to Jahoda's theory of the latent functions of employment and the effects of their deprivation. It is not just humans that will be affected by the rise of automation; the place of AI in society, particularly whether they will be included in society or not (questions of membership/citizenship), or whether they will be treated as second class citizens. I expand Michael Walzer's notions of 'hard' or 'unwanted' work, and how these concepts could inform the status of AI in society. Following this, I turn to Noreen Herzfeld's work on the Image of God and AI. Building on her argument that AI will image humans in ways similar to how humans bear the image of God, I attempt to expand her typology to include the transhumanism discussion. Using Herzfeld's expanded typology, I posit a number of questions regarding the status or needs of AI depending on how intelligence is defined and how enhanced humans or post-humans might also be affected. Finally, I apply the analysis of the health effects of unemployment to issues raised by transhumanism. Research has shown that participating in active, meaningful leisure activities can mitigate the deprivation of Jahoda's latent functions. I posit that increased religious engagement could therefore be an unexpected result of the rise of AI and transhumanism.

AI, automation, and unemployment

There has been considerable attention given to the potential loss of human employment through the development of AI and the automatization of work. In 2013, Oxford researches Carl Frey and Michael Osborne attempted to determine what percentage of jobs in the United States were susceptible to automation. Depending on the probability of this, they generated three categories, jobs that were at low, medium, and high risk of automation. It was their conclusion that 47% of jobs in the United States were in the category of high risk (Frey and Osborne, 2013, 44). The kinds of jobs in this category include transportation, administration, labor, and service work (44). They also found a negative relationship between wage, education, and automation, meaning that jobs with higher wages and that require higher levels of skill were less likely to subject to automation (5).

Subsequent research, however, shows that such estimates may in fact be too high. In March of 2018, the Organisation for Economic Cooperation and

Development (OECD) published a report that claims to build on Frey and Osborne's approach, with the results applying to the 32 countries in the OECD that have taken part in the Survey of Adult Skills (Nedelkoska and Quintini, 2018, 4). In this study, their findings do not seem that different with "close to one in two jobs are likely to be significantly affected by automation," but there is additional nuance to these findings (7). Nedelkoska and Quintini estimate that 14% of jobs in these 32 countries have a 70% probability of automatization, whereas 32% of jobs have a probability between 50% and 70%, of automatization (7). In the latter case, this could mean that some tasks are automated, not eliminating a job completely, but calling for new skills. This report also reinforces the 2013 findings that jobs that require the lowest levels of education are the most likely to be susceptible to automatization, and those that require more education are less susceptible (8).

There do appear to be differences in automation, loss of employment, and re-employment based on age. The study describes the relationship between age and automation as "U-shaped, but the peak in automatability among youth jobs is far more pronounced than the peak among senior workers" (Nedelkoska and Quintini, 2018, 8). Automation, therefore, is more likely to displace younger workers than older workers, but younger workers are more likely to be able to adapt to the technology or skills needed, so the transition from unemployment to re-employment is likely easier for younger works than older (8). However, research has made it clear that there are health impacts for both young and old based on the loss of employment.

Effects of unemployment

AI and automation have the potential to contribute to the further stratification of American society, including having a profound effect on those who find their work replaced by machines. A longitudinal survey was done with young Americans from 1979 to 1994, and using the results of these surveys, researchers examined a number of things, including whether there was a relationship between unemployment and symptoms of depression. After taking account of prior symptoms, socioeconomic status, and family background, the length of previous unemployment predicted symptoms of depression significantly, with the *p*-value being less than 0.01 (Mossakowski, 2009). The effect was more profound with men than women, with the author speculating that women might be stigmatized less because the expectation that they were in the workforce was lower. These results were taken from people between the age of 29 and 37 at the end of the study (1994), with the average period of unemployment being 1.47 years and the range from zero years to thirteen years out of the workforce. These results focus on younger workers who are unemployed for a relatively short time over the course of a career. It turns out that those who lose their jobs closer to retirement have greater changes in physical and mental health.

A study published in 2000 argued that the loss of employment for older workers has consequences not only for people's mental health but their physical health as well. Relying on data from the 1992 and 1994 Health and Retirement Survey, Gallo et al. restricted their 1992 baseline to respondents that had been employed for at least three years with their current employer (Gallo et al., 2000, S132). Physical and mental health were scored on scales of 0 to 15 and 0 to 8 respectively, with higher scores representing worse health (S133). The mean physical health score for those who did not experience job loss was 1.759, compared to 2.325 for those who experienced an involuntary job loss between 1992 and 1994. Additionally, those who experienced involuntary job loss also had higher rates of hypertension, obesity, cigarette consumption, cancer prevalence, and heart disease prevalence. The mean mental health score was 0.946 for those employed and 1.540 for those who experienced involuntary job loss. All of these findings were statistically significant, with a p-value of less than 0.05 (S137).

William Gallo, working with other researchers, published a study in 2006 that used data from the Health and Retirement Survey from 1992-2002 in order to assess the effect job loss might have on the risk of myocardial infarction and stroke (Gallo, 2006, 683). They found that individuals who experienced involuntary job loss have more than twice the risk of having a myocardial infarction or stroke compared to those who remained employed (686). In light of this, the researchers argue that doctors should identify the loss of a job as one nears retirement age as a known risk factor, particularly for vascular health. Additionally, policymakers should be aware, and include health consequences in the total cost of unemployment (686).

Latent functions of employment

Symptoms of depression are not the only problems those who are unemployed might face. A more comprehensive relationship between mental health and unemployment is Jahoda's theory of deprivation. The loss of employment, Jahoda posits, deprives individuals "of the latent functions that employment provides. These functions are time structure, regular shared experiences, information about personal identity, a link with collective purpose, and enforced activity" (Waters and Moore, 2002, 15). An Australian study builds upon Jahoda's work to examine how meaningful leisure activity might reduce both deprivation as well as mental distress in times of unemployment. Unlike the study involving symptoms of depression, this study engages adults, both unemployed and those working full time. Waters and Moore (2002) cite related research over the years that reinforces Jahoda's theory, including Evans and Haworth (1991), Haworth and Ducker (1991), and Feather and Bond (1983). These studies suggest that keeping busy during the day can help mitigate deprivation in young adults, and that active leisure activities increased fulfillment more than passive leisure activities among young adults. Finally, social leisure activities produced different effects than solitary activities.

Waters and Moore hypothesize that those who are unemployed admit to more deprivation, higher levels of depression, and lower self-esteem than those who are employed. They also believe that unemployed individuals engage in fewer leisure activities, and that they engage in fewer social activities than solitary activities and that they receive less meaning from these activities than employed individuals. Employed individuals are believed to engage in solitary leisure less than social leisure activities (Waters and Moore, 2002, 17). Unlike the U.S. study, Waters and Moore found no connection to the effects of unemployment and gender. Their findings suggest that employment status alone predicts sixty-five percent "of the difference in frequency of social and solitary leisure activities, meaning attained through social and solitary leisure activities, perceived deprivation of latent functions and psychological health" (20). It turned out that unemployed persons did engage in more solitary leisure activities and fewer social activities than those who were employed, though loss of income could also be a factor in this case. Unemployment did lead to increased deprivation of Jahoda's latent functions, and this was significant. The most significant of the latent functions was the deprivation of personal identity, which makes sense "if people think of themselves in terms of the job they do" (27). Meaningful leisure activity does reduce the distress of unemployment, and social leisure activities more so than solitary activities, which provides a place for constructive work regarding unemployment, identity, and mental health.

AI, hard work, and membership

On the other hand, AI and automation also provide the promise of saving human lives by replacing human workers in situations that are harmful to humans. Jobs that would expose people to harmful chemical, or dangerous working conditions could be done by machines who may not be susceptible to the environmental effects, or can be repaired or replaced more easily. While this could be positive for people, once the effects of unemployment are overcome, it does raise potential issues for machines, particularly intelligent ones. It has been shown above that humans make connections between their work and identity, but they also compare their value to others. There are jobs that are seen as undesirable, jobs that are believed to require no particular skills or education. These jobs confer little to no status socially, and these are the kinds of jobs that will most likely be done by nonhuman workers. The question then becomes, how do we treat these nonhuman workers, especially if they possess human-like intelligence?

Michael Walzer, in his book, *Spheres of Justice: A Defense of Pluralism and Equality*, examines how communities deal with the question of who is given or denied membership. Nations currently allow people to live within their borders who are not citizens, but Walzer ultimately believes the distinction between immigration and naturalization should be abolished. In other words, he does not see resident aliens as a positive solution, and would rather people who enter a country to live be given the full rights and privileges as citizens, while also agreeing to abide by the same set of rules that citizens do. This is not currently the case, though, and Walzer states that it is possible that "the state controls naturalization strictly, immigration only loosely. Immigrants become resident aliens and, except by special dispensation, nothing more. Why are they admitted? To free the citizens from hard and unpleasant work. Then the state is like a family with live-in servants" (Walzer, 1983, 52). He goes on to say that "the rule of citizens over non-citizens, of members over strangers, is probably the most common form of tyranny in human history" (62). Although Walzer is concerned about humans in the context of immigration, naturalization, guest workers, refugees, etc., I believe his concerns can extend to AI workers as well.

As I said earlier in this section, nonhuman workers will likely have to do the work that humans do not want to do. If this automation is done by machines or computers that are not intelligent, the concerns are far fewer. The comparisons made in that case would be with existing technology that makes work easier for humans, or at the most controversial, the use of animal labor. If, on the other hand, AI performs these tasks, I argue that Walzer's concerns about membership are relevant. If AI is comparable to human intelligence, then I think humans need to think about what place such workers should have in society. If their intelligence is like humans, will they also make similar associations between work and identity that humans do? I simply raise this question now, saving any attempt at an answer for the final section.

Imago Dei and intelligence: Herzfeld's typologies

Noreen Herzfeld, in her book, *In Our Image: Artificial Intelligence and the Human Spirit*, argues that the quest for AI says a great deal about our own nature. She takes up the questions: what characteristics of humanity do we value enough to want to image them in AI and what are the consequences of those choices (Herzfeld, 2002, 5). Herzfeld identifies three categories scholars think about related to humans being made in the image of God. These categories are described as substantive, functional, and relational. The substantive approach to the Image of God places the emphasis or the locus on an aspect of human reason. For Augustine it was the will; for Reinhold Niebuhr it was the capacity for self-transcendence. Regardless of the specific answer, those who fall into this approach believe there is something particular to humans that we only share with God. Critics of the substantive approach argue that there are few, if any, characteristics that humans do not share with other species. Self-consciousness is found in elephants, dolphins, and some primates. Genetically speaking, humans share over 98% of their DNA with their closest related species. Even if there is something unique to humans, the substantive approach is also criticized for being too anthropocentric.

Functional approaches to the Image of God place the locus in humanity's purpose. One such function that many scholars find support for in scripture looks to humanity's relationship with the rest of creation. Some view this relationship as one of dominion, that humans have been given creation to do with as they see fit, while others prefer the language of stewardship, that humans have a responsibility to care for creation as representatives of God (Herzfeld, 2002, 21). Critics of this approach argue that scripture does not provide a model for representing God that differs greatly from other civilizations of that time, many of whom only saw the king or ruler as having a relationship with the divine. Other critics could argue that humans have failed to live up to this function and therefore it cannot serve as the basis for the Image of God within humanity. The third approach, the relational interpretation, begins with the importance of relationality within God. The Trinity emphasizes the importance of relationality; it is a very part of the nature of God. Humans possess the same I-Thou relationship that God possesses, but some scholars extend relationality further (Herzfeld, 2002, 26). Barth argues that the creation of humans as male and female is an extension of the relationality found in the Triune God (Herzfeld, 2002, 26-27). Some criticize Barth because humans are far from the only species to have two sexes, while others argue that relationality is not the foundation of the Image of God, but rather a consequence of the Image, which is grounded either substantively or functionally (Herzfeld, 2002, 27-31).

Herzfeld finds a striking similarity between the ways in which intelligence is modeled in the context of AI and how theologians describe the image of God. Just as she provides three categories for the image of God, she therefore offers three similar categories for intelligence. Intelligence is seen as a quality, matching a function of human intelligence, or being able to relate to humans (Herzfeld, 2002, 34). Identifying the intelligence in AI as a quality, researchers believed that intelligence could be isolated and reproduced in mechanical structures. Referred to as symbolic AI, Herzfeld cites that this was the dominant approach on the field until the 1980s. The idea is that rational thought is made up of symbols, of which there are a fixed number used in combination with one another. Additionally, there are rules that govern the combinations and there must be a fixed number of these as well.

There is a divide in the AI community between strong AI and weak AI. Strong AI wants to replicate the full effects of the human mind in computers. Weak AI, on the other hand, focuses on one aspect of human intelligence, a particular function, and seeks to create AI that can do one task very well. Perhaps the most famous example of such a functional approach is Deep Blue, the chess-playing computer (Herzfeld, 2002, 43). It did not exhibit behavior that humans do in competition; it simply used the rules of the game and its superior ability to process moves ahead to overpower its human opponent.

The relational understanding of AI comes from the famous Turing Test. The abstract question of whether computers can think was impossible to answer without interacting with the computer. Alan Turing posited an experiment where a person is connected via computer to two subjects, and asks them questions, with the intent to determine which is a human and which is a machine. An AI that could fool the interrogator as often as it failed would "pass" the test and could be described as intelligent (Herzfeld, 2002, 45-47). Intelligence in this example, is only known in relationship. The way in which the AI is known to be intelligent is that it is able to engage in conversation with another, and be understood as a thinking other.

Jahoda's latent functions in the context of AI

Depending on which kind of AI one is talking about, questions can be raised as to how this kind of intelligence is similar to humans and how Jahoda's latent functions could apply to nonhuman workers. In this section, I examine each of Jahoda's latent functions using Herzfeld's typology of intelligence to raise questions about the status of AI.

Time structure

AI that is doing work will have enforced activity for sure, but will there be an understanding of structured time? If a machine is working nonstop is there really time structure? If intelligence is a substantive characteristic, and it is something that separates humans from other organisms, if AI possesses a substantive intellect as well, will they require time structure in the way we do? Will their minds require rest as ours do? If intelligence in AI is relational, should AI be given time to have social interactions, and since our intelligence is embodied, should we expect AI to be as well? Would a network connection be sufficient for AI socialization, or would it require physical presence as well?

Regular shared experiences

How do we define shared experience? For AI, would doing their own job, even if part of an assembly line be considered shared experience? If Deep Blue played chess with another Deep Blue, would they be sharing a chess match or competition, or simply acting out their designed purposes? If AI is relational, will they experience dissatisfaction if they are forced to work in isolation?

Information about personal identity

Will AI have the same relationship between employment and identity as humans do? It seems to me that functional AI could evolve a "strong" connection between the function they perform and a sense of identity. If AI is relational, is it possible for them to form a sense of personal identity without social interaction? If AI is substantive, will AI come to similar conclusions about themselves as humans do, given that intellect is one of the things that people associate with human identity?

A link with collective purpose

In what kind of collective purpose could AI engage? Is it possible to program collective purpose of AI as part of its overall purpose? For instance, as an intelligent agent, could the collective purpose of AI include the flourishing of intelligent life, or abiding by the categorical imperative, creating rules that apply universally to all rational agents? The rules commonly referred to as Asimov's laws of robotics focus on nonmaleficence, but the requirements of nonmaleficence are less stringent than the requirements of beneficence. If AI are programmed to act beneficently towards humans, or more generally to all intelligent agents, does this place a collective purpose within their own *telos*?

Enforced activity

Do humans have the right to enforce activity upon other intelligent agents? Even if AI is only functionally intelligent, is that sufficient for a preferred moral status? Enforced activity assumes that when one is not working, one has the choice to not be active. As stated in the time structure section, will AI be given the opportunity to have 'downtime', and if not, is enforced activity a latent function that helps shape identity, or is it a burden imposed on those powerless to argue against it?

Transhumanism and the possibility of second class citizens

I believe that the same arguments presented about membership in society and the specter of second class citizens surrounding AI and automation can also be made regarding transhumanism. Although there is significant diversity within the movement about how to improve humans, the overall goal is to take control of human evolution in order to move beyond the current limits of humanity. To my knowledge, there is not much discussion about forming their own society or leaving the planet in order to form their own society. In fact, there seems to be a somewhat naïve belief that everyone will want to take advantage of the improvements to humanity that they seek. Such universal embracement is unlikely, especially at first, so there will be a significant time when humans, enhanced or not, will exist together in society, potentially even with multiple post-humans species. This co-existence, however, presents the same problems as co-existing with AI.

Literature and popular culture provide numerous examples of how people envision the relationship between humans and beings with abilities surpassing humans. Though these imaginings cannot be understood as the eventual future, they do represent how people have already begun to think through the implications of the application of technology to human nature. Particularly when there are few enhanced individuals compared to humans, the relationship is often cast in terms of the dangers to humanity. Those who are not human, or beyond human are seen as possessing abilities that are too dangerous for individuals to possess. Although one could imagine a scenario in which transhumanists intentionally try and improve human moral capacities, it is difficult to see how that would currently be possible, and also hard to believe that every transhumanist would want to pursue such improvements.

On the other hand, the more accepted human modification becomes, the relationship between humans and those with enhanced abilities changes. In the 1997 film *Gattaca*, the protagonist is born into a world that has just started to utilize gene editing. His parents decide to have their first child trusting God, not their 'local geneticist', forgoing any improvements to his genome. It turns out he has a heart condition and the doctors say he will have a shorter life span than the average child. His younger sibling, born only a few years later, is conceived using IVF and gene editing to ensure a healthy embryo. By the time the protagonist is an adult, society has become stratified by genetic potential. Only those who have superior genetic profiles are eligible for the best jobs, whereas unaltered humans are relegated to lower-paying jobs. The main character illegally assumes the identity of an enhanced person who has been paralyzed, a practice that is stigmatized, with people referring to such pretenders as 'borrowed ladders' or 'de-gene-erates' (*Gattaca*, 1997). The enhanced person provides biological samples for the tests employees are subjected to in order to ensure they are who they say they are, and in turn receives housing and money.

If people are capable of increasing their intelligence far beyond that of current humans, they could argue that they know what is best for non-enhanced humans, enhanced humans, and post-human species. If they increase their strength sufficiently, they could argue they have the power and therefore should rule. If their lives are extended significantly beyond the human lifespan, they could argue policies have a longer-lasting effect on them, so they should have ultimate say, or by living longer, they can consolidate capital and political power over time. Such enhancements would give clear advantages to those who can currently afford access to the technologies used to enhance human capabilities. Without oversight, it is possible to increase the divides that currently exist in the world, potentially to the point where those with power and influence are so different from those who are not enhanced that they no longer see the inherent value in their lives.

Reimaging Herzfeld's typology for transhumanism

I believe that Herzfeld had an important insight when comparing discussions of the Image of God and models of intelligence in AI research. Though I do not think the comparison is quite as clear, in this section I attempt to apply the three-fold typology she uses to transhumanism. Instead of asking what aspect of human nature will be imparted to computers made in our image, the question in the context of transhumanism is what about human nature do transhumanists seek to improve or replace? The first category of the typology is the most straightforward, substantive enhancements. These are specific traits or characteristics that humans possess that transhumanists want to enhance, increasing the depth or power of the capacity. It could also apply to characteristics that transhumanists want to be incorporated into a new or post-human nature that humans do not currently possess.

The second category of the typology, the functional approach, is probably the most difficult of the three to apply to transhumanism. Functional here is related closer to the typology of the Image of God than it is intelligence in the context of AI. Though it is plausible that transhumanists would still want humans and post-humans to steward creation, I would argue the primary function transhumanists value is the role of taking greater control of human evolution. The motivations behind this role might differ, such as a desire for immortality, or to minimize the role of chance in evolution and the removal of harmful conditions from the human genome, but what agents are doing remains the same; intentional, direct changes to what is considered human with the purpose of improving the human condition, or creating a species that is an improvement.

The third category, the relational aspect, focuses on how humans or post-humans relate to their bodies and to the rest of creation. Although some might view this as a subset of substantive enhancements, I argue that the emphasis here is on how individuals exist in relationship. Reinhold Niebuhr referred to the perfection of the creaturely aspect of human nature as fulfilling the natural law (Niebuhr, 2004, 270), which he defines as harmony within three sets of relationships: one's relationship with God, an ordered soul (relationship of the self within the self), and one's relationship with the rest of creation (Niebuhr, 2004, 288-289). Transhumanism offers the possibility to improve all three of these sets of relationships. Increasing empathy or enhancing spiritual experiences could bring one closer to God. Tattoos that identify changes in blood sugar, or instant biofeedback that one can use to control breathing, meditation, blood pressure, etc. improve one's relationship with one's body and self. Expanding the range of light that people can see, or enhancing other senses can make people more aware of how they impact the world around them.

Jahoda's latent functions in the context of transhumanism

Much like with AI, how Jahoda's latent functions might interact with transhumanism depends on the specifics, such as the type of enhancement or the scope of transformation. I do not present an exhaustive list, but rather try to identify issues relevant to transhumanism that arise when reflecting on the expansion of Herzfeld's typology and Jahoda's latent functions. One question that could be asked of each of the latent functions is that if substantive transhumanist changes are pursued, such as increased intelligence, will this make enhanced humans or post-humans even more susceptible to mental distress due to deprivation of these functions?

Time structure

If radical life extension is achieved, how might agents reimagine time and how it is structured? If one has hundreds or thousands of years to live, will this decrease the importance of structured time, because there will always be more time to accomplish something, or will the specter of eternity make people crave more structure and order to their lives? If people are living longer, will they become less short-sighted in terms of pursuing ends that lead to instant gratification?

Regular shared experiences

If people are allowed to pursue a wide variety of enhancements, genetic or mechanical, and augment the ways in which they experience the world around them, it is possible that experiences will become less and less shared the more customized one becomes. Or will augmentation, enhancement, and transformation in general provide the underlying shared experienced regardless of what kinds of changes one chooses to make? Will longer lives lead to greater shared experiences as people spend more of their lives in work, or pursuing meaningful activities? If so, could this lead to a stratification of society, increased nationalism, or vilification of those not in one's in-group on a scale not seen before in human history?

Information about personal identity

A functional understanding of transhumanism might lead to an increased sense of personal identity. If humans are to co-create or take control of human evolution, this could provide new insights for people as to their identity and purpose, especially with the existing connection between work and identity. On the other hand, if people make many substantive changes, people could claim that changing a number of one's traits may raise fundamental questions about what one's identity is. Just as there is the possibility for shared experience to fracture, personal identity could become a nagging question, or it could be tied far too closely to the kinds of enhancements one chooses, a form of reductionism.

A link with collective purpose

The questions raised in the previous two sections are relevant here as well. Will collective purpose be strengthened as common goals are pursued, or will purpose become fragmented and divided? If people wish to engage in work to minimize the effects of AI and transhumanism on people's employment and mental well-being, then I would think a strong sense of collective purpose would remain. The same would be true if intelligent agents work together to save the planet from environmental disaster. Collective purpose could run in parallel, with the collective purpose of everyone is to pursue their interests to the betterment of society as a whole, so people may not be working together or have shared experience, but are pursuing the same kinds of ends in a variety of ways.

Enforced activity

Radical life extension raises questions about enforced activity in both directions. It would stand that with longer lives, there should be longer periods of time in which people can choose what kind of activities they wish to pursue. On the other hand, if overpopulation is a concern and people have fewer children, the next generation will not exist in sufficient numbers for people to stop working. Retirement could be much farther off than people imagine. Of course, AI and automation could alleviate this problem, but there will always be a need for a certain amount of a fixed population to do certain kinds of work. Will those with enhanced strength compel the weak to work? Will the intellectually enhanced attempt to make decisions for everyone, claiming a superior understanding of how things work and the consequences of their decisions?

Addressing concerns thus far

Before turning to the issue of religious engagement, I think it is important to revisit some of the issued raised thus far and provide a way forward or tentative solutions. As I stated in the section on collective purpose, I believe that the *telos* of AI needs to include the flourishing of humans as well. A possible way of doing this is to move from nonmaleficent language to the more stringent standards of beneficence towards humans, or intelligent agents in general. Just as the perfection of human nature for Niebuhr can include how humans relate to others, which is another aspect of human nature besides intelligence that could be given to AI.

If the relationship between humans and AI is of primary importance to AI, humans will also need to prioritize the relationship, or risk AI questioning why the relationship is predominately one way. Depending on how one views the intelligence of AI, different ways of dealing with this situation exist. If AI is functionally intelligent, humans could still decide to treat AI the same way other intelligent beings are treated. If intelligence is substantive, then AI should not be forced to work nonstop, and be given the opportunity to rest and reflect like human agents do. And if intelligence is only identified in relationship, then AI must be given the space and time to interact with others in ways that are meaningful to them.

In order to keep AI from being treated as second class citizens, the previous paragraphs comments are a start, but are insufficient. If AI are performing the kind of work that humans or post-humans do not desire to do, then they must be given membership in society and have a voice in collective decision making. Although a subservient class of automated, emotionless workers might be appealing to some, human intellect is rarely experienced as completely disconnected from emotion. With the rise of AI and automation, steps can be taken to ensure that unemployed humans do not become second class citizens either. The first step is to work to reduce the stigmatization of unemployment. There will be many skilled workers who will be replaced by nonhuman workers; it has more to do with safety or finances than it does the worth of individual employees. A significant increase in unemployment will also likely require government action. This could include forming a universal basic income, or legislating the regulation of automation with taxes or charges that benefit directly those whose jobs have been displaced.

There are multiple ways to ensure that neither enhanced humans, post-humans, nor humans become second class citizens. The widest-reaching of these would be to legislate genetic freedom, with caveats to public welfare and health. If technological modifications are safe, people should be free to choose whether or not to pursue particular enhancements, and not be penalized in the workplace of society for their choices. If there are concerns about the expense of enhancements, then one way to level the playing field would be to institute a lottery. Anyone seeking a particular enhancement could submit their name, and then people are randomly chosen to receive the enhancement at a sliding scale of cost. If the enhancements are meant to improve society or the species as a whole, then a lottery is one way to minimize the chances of a stratified society based on the enhanced haves and the nonenhanced have nots. Human, enhanced humans, and post-humans all need governmental representation if they all coexist in society. In order to help agents deal with the stressors of unemployment, additional opportunities need to be made for people who are unemployed to engage in meaningful social leisure activities. This could include local governments creating volunteer projects, sports leagues, expanding representative governance, or as the final section of this paper argues, religious engagement.

Religious engagement

As stated earlier in the paper, social, active, and meaningful leisure activities provide access to Jahoda's latent functions when people are deprived of them through unemployment. I would argue that religion fits the definition of a social, active, and meaningful leisure activity and that it is possible that the rise of AI and transhumanism could lead to renewed religious engagement. First, religion is a leisure activity, that is, something done in one's free time. Except for those who are paid for their professional contributions, those involved choose to do so, not as part of a job. Although there are aspects of religion that can be done by individuals, by its nature it is social, involving a community of people with shared beliefs who also interact with their local community and to some extent, the world. Religions are at their best when they are active. They have people to come together and participate in worship and service. Because religion deals with questions of ultimate reality or ultimate concern, it is also meaningful. Classifying religion as a leisure activity is not meant to be a criticism of it, nor is it meant to diminish the impact it can have on people's lives. In the context of this paper, a leisure activity is merely a categorical distinction from employment. People who are not paid to work for a religious institution voluntarily decide to pursue religion in their free time, thus it is a leisure activity. Though the details of religious activity may differ from a kickball league or volunteer work at an animal shelter, these activities all fall into the same sociological category of leisure activities.

Time structure

Religion structures time in several ways. It provides a time separation between the sacred and the profane, as well as structuring time within the sacred portion of time. Liturgy divides corporate worship into distinct periods, where music, spoken word, or prayer are the focus. Outside of corporate worship, there are often many opportunities for people to take part in activities related to the religious community, such as classes, fellowship, service, music rehearsals, and meals. For those who are involved deeply or regularly, these commitments can shape not only the day in which corporate worship takes place, but the entire week, restricting involvement in other leisure activities.

Regular shared experiences

There are weekly or sometimes daily chances for corporate worship. People are encouraged to read sacred texts at the same time, and meals often come before or after worship. Even if one cannot physically make it to the place of worship, many organizations offer live or recorded broadcasts on radio or the internet. Those who participate in corporate worship hear the same sacred texts being read in a given service and sing the same songs as well, regardless of their neighborhoods, employment (or lack thereof), education level, or income level.

Information about personal identity

Religion also provides insight into one's place in the universe and questions of personal identity. People are placed in a history with those who share the faith as well.

A link with collective purpose

A local community of faith has its own purposes and missions, but there is also a connection with those around the world who share the same faith. Worship may be one aspect of the collective purpose, but action in the world is also a part of religion's purpose.

Enforced activity

Religion requires something of its participants. It can be attendance, belief, financial giving, renouncement of particular behaviors or material goods, meditation, prayer, reading, time, or any combination of these and additional requirements.

Religious engagement in light of AI and transhumanism

If AI has intelligence that is similar to humans, it is possible that religion could also be an activity that could reduce mental distress if they are being deprived of Jahoda's latent functions. If a machine does not feel a sense of collective purpose from its work, it may choose to supplement its activity to include religion in order to have that need met. Post-humans with enhanced intelligence may not need the information about personal identity that religion offers, but religion might provide a structure for time or shared experience that is desperately needed. A machine whose time is strictly structured might not have a sense of identity tied to its work. A future version of Deep Blue with human-like intelligence may not feel that chess comprises its ultimate concern and therefore could use religion for additional information about being an intelligent agent in the world.

So far, the examples I have raised assume that AI, enhanced humans, and post-human species have more similarities to humans than not. It is possible that even if Jahoda's latent functions are operative, which they may not be in other forms of intelligent life, that religion could also be altered radically. Just because religious engagement can help with mitigating mental distress does not mean it will be the kind of religion that has existed previously. Already people are attempting to create AI to be worshiped as a form of superintelligence (Harris et al., 2017). The Way of the Future is a Silicon Valley-based church that seeks to create divine AI. Anthony Levandowski started the organization as a non-profit and chose to envision the project as a church rather than a company to encourage participation and to defend against claims of doing this for profit (Harris et al., 2017). If enhanced humans relate to the world in fundamentally different ways than modern humans, perhaps different wavelengths, temperatures, or chemical compounds will be used to bring about new ways of experiencing the spiritual. Even the mythology of religions could change. AI might not identify as being made in the image of humans and may assert their own myths typing them into the tradition of *Imago Dei*. Post-humans who decided on their own gene editing may reject any notion of the image of God. There is much uncertainty as to how these technologies will develop and what influence they will have on humans.

Conclusion

In order to attempt to bring the different strands of conversation in this chapter together, I end with some suggestions for things to take into consideration or for directions for future research. The first set of suggestions surround the intersection of employment and automation. I agree with the assertion of Gallo regarding the acknowledgment of late-career job loss as a medical risk factor, that as automation is planned there needs to be parallel planning to help retrain workers for re-employment, and that the health consequences of loss of employment need to be taken into account when examining the impact of unemployment. If automation saves money for businesses, but in doing so harms people's economics and health, then perhaps automation should be taxed to bear some of that burden. This idea has also been proposed by Bill Gates, who argues that this will slow down the process of automation and could contribute funds to jobs that are not well automated, such as care of the elderly and the young (Delaney, 2017).

The second set of suggestions deal point to proactive steps that could be taken to deal with the health consequences of jobs lost to automation. Knowing that there are major health consequences to involuntary job loss and that there are things that can mitigate these, such as participation in social, active, and meaningful leisure activities, a preventative approach would include incentivizing people to take part in these kinds of leisure activities while employed. Waters and Moore showed that those who have experienced job loss are less likely to engage in active, social, and meaningful leisure activities, so if people are already participating while they are employed, it is possible they will continue to do so, lessening the deprivation of the latent functions of employment. It would also be helpful to find out active, social, and meaningful leisure activities have any impact on the physical health of those who have lost jobs, particularly those who are older. The research thus far has shown these leisure activities can help with the mental health of people, but further work is needed to know whether they also can mitigate the negative impact on physical health.

Finally, a third set of suggestions will be made in the context of AI, transhumanism, and religion. AI and human enhancement, as stated earlier, could lead to renewed religious engagement, or even new forms of religious expression. Eventually, the intelligence or power of AI or enhanced humans or post-human species could be so great that non-enhanced humors could imagine them to be omnipotent or omniscient. Traditional religion would see such adoration of material persons as idolatry. Mark Graves, in the discussion of machine learning and ethics, posits three values to help shape relations between humans and human-level machine intelligence (HLMI) that would also be useful in this discussion. These values state that both entities: (1) "have the right to exist and to hold values to which they ascribe. (2) Neither HLMI nor humans should thwart the right of others to commit to their values, even those which may conflict with one's own, as long as those values do not conflict with the right of others to hold values. (3) Human and HLMI should communicate and attempt to understand each other's values regardless of whether one holds or identifies with the other's values" (Graves, 2017, 342-343). Another way to potentially inform AI about the nature of human religion is to give it data; machine learning could be used to analyze sacred texts and the works of great religious thinkers, providing concrete examples of what is valued. AI might even be able to make connections between different schools of thought that humans were unable to reconcile on their own. Love is at the core of many if not all religions, so perhaps the best way to express that value is to extend it to other beings, whether they be unemployed, mechanical, enhanced, or a new species. If AI witnesses how people are dealt with as their work is replaced by machines, they might imagine this is how they could be treated in time. Avoiding second-class citizens in any form could be the greatest religious expression to shape how intelligent agents should interact with one another.

References

Delaney, K., 2017. The robot that takes your job should pay taxes, says Bill Gates. *Quartz* [online]. Available at: <https://qz.com/911968/bill-gates-the-robot-that-takes-your-job-should-pay-taxes/> [Accessed 10 June 2018].

Frey, C. and M. Osborne., 2013. The future of employment: how susceptible are jobs to computerisation? Available at: <https://www.oxfordmartin.ox.ac.uk/downloads/academic/The_Future_of_Employment.pdf> [Accessed 7 June 2018].

Gallo, W., 2006. The impact of late career job loss on myocardial infarction and stroke: A 10 year follow up using the health and retirement survey. *Occupational and Environmental Medicine*, 63(10), pp. 683-687.

Gallo, W., E. Bradley, M. Siegel, and S. Kasl., 2000. Health effects of involuntary job loss among older workers: findings from the health and retirement survey. *The Journals of Gerontology Series B: Psychological Sciences and Social Sciences*, 55(3), pp. S131-S140.

Gattaca, 1997. [DVD] Directed by A. Niccol. Hollywood: Columbia Pictures.

Graves, M., 2017. Shared moral and spiritual development among human persons and artificially intelligent agents. *Theology and Science*, 15(3), pp. 333-351.

Harris, M. et. al., 2017. Inside the first church of artificial intelligence: Backchannel. *Wired* [online]. Available at: < https://www.wired.com/story/anthony-levandowski-artificial-intelligence-religion/> [Accessed 3 July 2018].

Herzfeld, N., 2002. *In our image: Artificial intelligence and the human spirit.* Minneapolis, MN: Fortress Press.

Mossakowski, K., 2009. The influence of past unemployment duration on symptoms of depression among young women and men in the united states. *American Journal of Public Health*, 99(10), pp. 1826-1832.

Nedelkoska, L. and G. Quintini., 2018. Automation, skills use and training. Available at: <https://www.oecd-ilibrary.org/docserver/2e2f4eea-en.pdf> [Accessed 3 August 2018].

Niebuhr, R., 2004. *The nature and destiny of man.* Louisville, Ky: Westminster John Knox Press.

Walzer, M., 1983. *Spheres of justice*. New York: Basic Books.

Waters, L. and K. Moore., 2002. Reducing latent deprivation during unemployment: The role of meaningful leisure activity. *Journal of Occupational and Organizational Psychology*, 75(1), pp. 15-32.

11.

Three theologies that influence how we view AI, technology, and the world

Christopher J. Benek

The CoCreators Network

Abstract

Many science and tech experts like Stephen Hawking and Elon Musk have warned of potential dangers of Artificial Intelligence. They have widely voiced dystopian interpretations of the future. In contrast to such dystopian views of the future is the emergence of progressing interactions and iterations of theology and technology. AI is forcing theologians to reconsider humanity's previous concepts of self and humanity's role in the cosmos. These considerations are, in turn, forcing technologists to consider the value of religion as it pertains to the development of AI. As a result of these changes in thought, both parties are now being forced to ponder if theology and technology are bound for convergence via human enhancement. Such theological and technological considerations, presented in this chapter, will radically impact our understanding of the function of technology, our concepts of being, and our notions of human purpose in the world.

Keywords: Artificial Intelligence, Technology, Theology, Supremacy, Escapism, Participatory, Identity

Introduction

Artificial Intelligence is forcing theologians to reconsider humanity's previous concepts of the self and humanity's role in the cosmos. In turn, these considerations are compelling technologists to consider the value of religion as it pertains to the development of AI. These changes in thought are challenging both parties to ponder if theology and technology are bound for convergence via human enhancement. I contend that such theological and technological considerations will radically impact our understanding of the function of technology, our concepts of being, and our notions of human purpose in the world.

The initial challenge in understanding how theology and technology are impacting one another requires one to consider the deficiencies of modern Western theology and its influence on culture. As this work will establish, important aspects of modern theology's narratives have been fabricated as being Biblical. Subsequently, the majority of the modern American church and modern culture have traded the Gospel for variations on a theological theme based on what equates to theological tall-tales.

The cartoon *Two Scientists Talking While A Scared Robot Hides* (Flake, 2016), by *New Yorker* cartoonist Emily Flake, expresses this theological predicament. The cartoon, whose image is precisely reflected by its title, is accompanied by the caption: "I don't know – the minute I figured out how to make it self-aware it realized that it was naked, ran, and hid from me." The joke here is found in its play on, Genesis 3:7, the commonly known Biblical passage wherein, the first man and woman made by God, have their eyes 'opened' and consequently they recognize that they are naked.

If we put the well-intentioned attempt at humor aside for a moment and actually read the Biblical narrative, we quickly recognize that the cartoon is in reference to a misinterpretation of what transpired in the Biblical story. The man and woman in the garden, as depicted in Genesis 3, did not hide in shame because their Creator made them self-aware. Rather, in the Biblical text, the role that the humans play in relationship to the robot in the cartoon is parallel to the role that the serpent plays toward the man and woman in Genesis. The man and woman's shame arose out of separation from their Creator as a result of their actions (The Bible, Genesis 3:2-7). Through their lack of self-limiting, they chose to live outside their divine purpose as caretakers of Creation. Thus, per the robot cartoon, there is a significant difference between humanity emulating God and emulating the serpent.

Unfortunately, nuanced theological differences have been mostly overlooked in culture, particularly in the field of technological development because the church itself has perpetuated theology that is not consistent with the Biblical narrative. Additionally, pastors have not done the challenging work of widely propagating an updated technological hermeneutic through which to understand the emerging technological world that we live in today. As a result, our old paradigms have failed to meet the theological needs of the modern scientific and/or technological individual.

Fortunately, in 2009 Pittsburgh Theological Seminary (PTS), in collaboration with the American Association for the Advancement of Science, recognized this need in the larger Christian community and developed the world's first Doctor of Ministry (DMin) program focusing on the intersection of Science and Religion. They understood that many scientists and technologists felt required to check their proverbial science-card at the door of the church. PTS hoped that educating pastors to be broadly fluent in fields of science and technology would provide them with the tools required to begin contextually addressing questions of ultimate meaning in the lives of scientists and technologists.

As a result, the seminary gathered a group of mostly second-career pastors who had been in scientific or technological fields in a prior season of life. As the youngest member of the seminary's first DMin cohort, I chose to use my final project as a way to try and understand how congregants in my local church viewed the intersection of theology and technology. The results were fascinating.

My project surveyed congregants of varying demographics to better understand their perspectives before and after a series of classes that featured a combination of theological and technological investigation and education. First, congregants, who served as the students for the class, were introduced to concepts of participatory theology presented through a reading and discussion of N.T. Wright's book *Simply Christian* (2006).

Wright's work is divided into three main sections. In the first section he attributes four characteristics of humanity—our desire for justice, a thirst for spirituality, the longing for relationships, and the attraction of beauty. He claims that these elements of our being point toward God. In the second part of the book Wright leads the reader through a conversational exploration of Christian doctrine pertaining to God, Israel, Jesus and the Coming of God's Kingdom, Jesus: Rescue and Renewal, God's Breath of Life, and Living by the Spirit. He uses this survey as a way of articulating that God is neither identical to creation nor apart from it—but instead is actively involved in the universe in a way that causes the increasing "overlapping and intersecting" of heaven and earth. According to Wright (2006), the end result of this overlapping and intersecting process is the increased 'interlocking' of heaven and earth, which, once completed, will fulfill the redemptive process that God has initiated through Jesus Christ. In part three Wright proceeds to illustrate then how humanity can participate in Christ's redemptive purposes in the universe. He says that this can be accomplished by being actively engaged with worship, prayer, Scripture, the sacraments, and by continuing to formationally develop the four characteristics outlined in section one of the book by learning to follow the teachings of Jesus Christ.

After the students gained an understanding of participatory theology, they viewed *Transcendent Man: Prepare to Evolve* (Kurzweil, 2009), a documentary about inventor, futurist, and author Ray Kurzweil. In the documentary, Kurzweil discusses his thoughts on technological development. He specifically predicts how advancements in artificial intelligence, genetics, nanotechnology, and robotics will result in the development of a human-machine civilization.

The film also includes the voices of respected experts in these fields and conveys their views on Kurzweil and their projections regarding the future of technology and how it will impact humanity. Admittedly, many of my congregants articulated that learning about emerging technology through *Transcendent Man* was like 'drinking from a fire hose' yet, they widely expressed that recognition of emerging tech will be a crucial part of humanity's developing future. One of the major concerns conveyed was about what are the intentions of people who were presently developing this technology. Finally, my students recognized that, as Christians, they ultimately had a divine responsibility to help steward such technology for God, though admitting that they did not know what such stewardship might look like in practice (Benek, 2014b).

As a result of these findings, I personally questioned why more Christians do not actively recognize that they too had this divine calling that my congregants had self-identified. Through significant study and practical exploration, I discovered a possible elucidation to my question. Mainly, that there are three competing theological models that dominate the modern church and culture.

Supremacy theology

The first of these three models is what I call *Supremacy Theology*. Supremacy Theology is a theological model that is practically enacted when its adherents refuse to compromise any theological position they hold. Supremacy Theology arises out of hubris. It is the classic 'my way or the highway' position applied to theology and extended to any or all other elements of life. The extension readily reaches into our technological views and practice as well.

In Christianity, God is the God of possibility (Matthew 19:26). God is not limited to our human options so when we think that the only way is *our* way, then we limit God. In this way, Supremacy Theology predisposes people to anxiety. We worry that if we do not do the task before us in our specific way, then things will go wrong.

The methodology behind Supremacy Theology can most commonly be seen enacted presently in American politics. In the example of the last presidential campaign in the United States, it did not appear to matter whether or not you supported Donald Trump or Hillary Clinton, people enacting the principles of supremacy theology would affirm that the opposing candidate was incapable of making any good points. The methodology behind Supremacy Theology enacted in American politics states that my candidate is always right and your candidate is always wrong.

Unfortunately, Supremacy Theology does not afford the opportunity for compromise or reconciliation. Every interaction is an engagement from a position of conflict methodology. Every interaction is a zero-sum game.

This is not to say that there are not times when it is valuable to uphold one's principles, moral values, and virtues. But we often outright miss the mark when we fail to properly consider positions and considerations of context and community. When being 'right' and winning becomes the final goal of interpersonal interaction with others we have already forsaken anything rightly resembling principles, moral values, and virtues.

Supremacy Theology is evident in the day-to-day life of Christian communities when we marginalize or silence voices that make up the collective body of the church. Supremacy Theology is present when we compete rather than collaborate with one another to create new ways forward. Supremacy Theology is made manifest in the church when people disengage from the accountability of a faith community because they want to chase their own pursuits. Supremacy Theology is apparent when, instead of humbling ourselves to find a better way, we coerce, wear-down, or lobby for our own way or hermeneutic instead of seeking to discover and emulate Christ's way.

Certainly the methodology undergirding Supremacy Theology is also prevalent in many facets of technological development and implementation. We see such methodology in the lack of accountability in many social media platforms. Such methodologies are further exacerbated by the development of code, fake news, and or spam bots that are created as ways to inordinately sway public perspective and opinion. In turn, such propaganda stifles well-intended community conversation and discernment.

Perhaps the greatest threat to humanity resulting from the methodology behind Supremacy Theology comes via the secular transhumanist goal of self-actualization that follows along the lines of the Nietzschean philosophy of *der Will zur Macht*. When technological advancement is solely aimed at favoring the individual we marginalize the rest of humanity and risk enslavement (Benek, 2014a). Supremacy Theology poses a considerable risk to the future of humanity.

Escapism Theology

The second of the three competing theological models is *Escapism Theology*. Escapism Theology is a theological model that ultimately encourages its adherents to avoid or leave behind their problems instead of facing them. It is a position that has been perpetuated widely by the American church in the mid-1900s.

The general theological position of Escapism Theology is that the Earth is going "to Hell in a handbasket" so all we need to do is say that we believe in Jesus as our Lord and Savior and then wait for Jesus to come back so we can leave earth behind and go to Heaven. I believe that it is this mentality that is the reason for the decline in the Western Church. Escapism Theology promotes apathy in people.

There is a great cartoon in *The Funny Times* by Raymond Lesser (1998) that illustrates this problem with Escapism Theology. It has two people sitting on clouds in Heaven—apparently doing nothing—and the caption says: "I miss stress." The cartoon depicts the way culture often depicts heaven—a place where we go to sit on a cloud and play the harp for eternity as a way to escape our present reality. Not only does that sound like a really boring eternity but it is certainly not a Biblical view of heaven.

Escapism Theology has come to prominence through a misunderstanding of the concepts of the reformation phrase 'by grace alone'. Escapism Theology has propagated the falsehood that, ultimately, there is nothing that humanity can do to contribute to the saving of the world. Again this is simply not a Biblical understanding of humanity or the cosmos. When humans really experience the grace of God, it moves us to respond.

The most practical metaphor I can think of to give to explain this necessitated response is to use an example of my two-year-old son. My wife and I took our son to the Magic Kingdom in Orlando for a day. My son was very excited to see Mickey Mouse for the very first time. We stood along the side of the road and I held him up so he could get a better view as we watched the *Festival of Fantasy Parade*. If you have not seen that parade, it is really pretty unbelievable— simply a spectacular sight.

My son saw all the characters and the costumes and the dancing and heard the music and his eyes were huge; he was taking it all in as a human being who is relatively new to the world. At the end of the parade, he saw Mickey and Minnie and he waved with excitement. After it was all over I asked him, "Did you like that?" He looked me right in the eyes, kissed me on the mouth, and then threw his arms around me and hugged me. This is similar to a grateful person's response to recognizing God's grace. When we experience God's grace in our lives, we can't help but show our love and adoration.

On the contrary, Escapism Theology tends to have an element of entitlement based in works righteousness. This is ironic because the non-Biblical premise underlying Escapism Theology is that, because of God's grace, all we really need to do is confess and believe in Jesus and eventually we will leave this "evil" world behind and go to heaven. One of the problems with this belief though is that it leads to transactional theology. If I do 'X' then 'Y' will happen—which is, by definition, works righteousness. When we ascribe to escapist notions our public witness to 'accept Jesus as my Lord and Savior' is effectively reduced to us bargaining with God to 'save us a spot'. Such theology is nothing more than a works-righteousness hangover disguised as grace.

An experience of true grace means that there is *no action* that we can engage in to earn the grace we receive. The gift is presented to us freely, not as a transaction, regardless of whether we choose to accept it or not. But if we receive this gift, our response does not and cannot become a payment for the gift because grace is a gift—it cannot be bought. Instead, as those do who are recipients of good gifts we are moved in love to express our gratitude, our joy, and our reciprocal love in response.

This understanding and proclamation of grace have failed to be communicated in much of the modern church. Similarly, the result in much of Western, Capitalistic culture is that our relationship with God has become transactional instead of adoring. It is as if the majority of Christianity has become stuck in the third stage of grief (bargaining) because we have widely failed to genuinely know God and make God known.

A consequence of a perpetuated transactional and bargaining mindset, as it pertains to understanding humanity's relationship to God or addressing questions of ultimate meaning, is that it forms people in ways that tend to avoid tackling problems directly. While the Biblical model of conflict resolution does encourage people to retreat at times so that they can advance, it does not advocate the avoidance of addressing evil. To the contrary, we are to intentionally tackle problems with God's help.

Unfortunately, notions of Escapism Theology have been popularized in the modern era by some of the world's finest Christian leaders. Most worthy of note was Billy Graham, widely known and is frequently placed among the most influential Christian leaders of the 20th century. According to his website, Graham preached to live audiences of more than 210 million people in 185 countries and territories over the course of his ministry (Finstuen, 2017). He provided spiritual counsel to 12 US presidents and to the Queen of England. He was an advocate for many new ministry developments, among them prominent TV Evangelist shows like Robert H. Schuller's *Hour of Power* telecast that at one point attracted 1.3 million viewers in 156 countries (The Star Staff, 2010).

Unbelievably, such grandiose examples just skim the surface of Graham's influence on humanity and culture. He wrote 33 books, many of them bestsellers. He started the weekly *Hour of Decision* radio program heard around the world. He had a syndicated newspaper column and instigated the founding of several magazines. He was listed 61 times by the Gallup organization as one of the "Ten Most Admired Men in the World." He appeared on the covers of *Time, Newsweek, Life, U.S. News and World Report* and numerous other magazines (Billy Graham Evangelistic Association Staff, 2018). In 1989 Graham became the only minister, functioning in that capacity, to receive a star on the Hollywood Walk of Fame (Stolberg, 1989). Lists of his accomplishments and the people he influenced directly throughout the world go on and on.

As it pertains to Escapism Theology, Rev. Graham was often quoted as saying "My home is in heaven. I'm just passing through this world." He is renowned for statements like: "The Bible says that as long as we are here on Earth, we are strangers in a foreign land. There are enemies to be conquered before we return home. This world is not our home; our citizenship is in heaven." The difficulty with quotes like these is that they are emotionally compelling and provide great media sound bites, but they do not offer the bigger picture of the Biblical message.

Generally speaking, most people would not question Rev. Graham's ability to lead people to Christ or his tremendous value to humanity. But the Christian Tradition has long articulated that God is an agent of reformation and redemption and the church—comprised of called people—is the object and catalyst of God's reforming and redeeming work in Jesus. As such it is pretty straightforward to believe that Christians, to whom folks like Rev. Graham have faithfully passed on the mantle of Christian stewardship, have a responsibility to evaluate how the church is doing presently in order to continually strive to make things better via the grace of God. A quick assessment makes it largely clear that things are not going well with the institutional church in the United States and Europe.

When people have been led to believe that all they have to do is 'accept Jesus as their Lord and Savior' and then they are 'good to go' to heaven, it becomes problematic because this kind of thinking is not Biblical. Moreover, because Christians have perpetuated unbiblical theology for decades—the church has effectively become spiritually impotent.

Additionally, Escapism Theology is its incomplete expression of humanity's future. Yes, the Bible says that there is a place called Heaven but its existence does not mean that humanity is called to flee the Earth. Instead, we are called to participate in the redemptive purposes of Jesus to help redeem and renew the world, in Jesus' own words, 'on Earth as it is in Heaven'. From a Biblical perspective, the end game of our participation in these redemptive purposes is that it will bring about the creation of a new Heaven and a new Earth in Christ, intersect and interlock with one another.

Why does this matter? Because it means that instead of escaping from the world humans are called to engage it. Humanity's divine calling is literally to strive to make the world and all of humanity better with God's help. That means that people have to choose over and over, day-by-day, hour-by-hour, minute-by-minute, in every context, to open the gift of God-given grace to walk in the formational teachings of Jesus. When we fall prey to Escapism Theology it opens up a bunch of excuses for humanity. It permits Christians to theologically ignore how we care for the Earth and implies that because humanity will not always stay here, it is acceptable to pollute the atmosphere, pour oil and dump plastic into the oceans, cut down the rainforests, and bring other forms of life to extinction because this isn't our long term home 'we're just passing through'.

As a result of Graham's influence, albeit potentially altogether unintentional, Escapism Theology has flourished. But, while Graham was prolific in his ministry, this wasn't the only way Escapism Theology was disseminated globally. In Christian culture such escapism theology arose within North American evangelicalism. In many of those circles the word 'rapture' is used as an eschatological word based on a version of premillennialist dispensationalism that depicts an end time when all Christians will rise into Heaven and join Christ while non-believers will be 'left behind' on Earth for an extended tribulation time (McKim, 2014; Chopra, 2005).

While there are a variety of specific theological speculations as to how such end times events unfold in premillennialism dispensationalism, such pre-tribulation rapture theology, was foundational in the formation of the wide breadth of theological views that are encapsulated under the umbrella of modern Escapism Theology. Originating with eighteenth-century preachers such as Cotton and Increase Mather such ideas were popularized well before Graham in the United States by John Nelson Darby, the Plymouth Brethren, William E. Blackstone and the Scofield Reference Bible (Markham, 2013; Olson, 2003; Blackstone,1908; Magnum and Sweetnam, 2009). Additionally, Robert Maton, Nathaniel Holmes, John Browne, Thomas Vincent, Henry Danvers, William Sherwin, Philip Doddridge, John Gill, Edward Irving all had early expositions and manifestations of the concept of rapture, and consequently Escapism Theology, in their works (Watson, 2015; Doddridge, 1738; Gill, 1748).

Even prior to these individuals, the Scottish minister, Edward Irving was dabbling in concepts in connection with premillennialism/Escapism Theology thinking (Tregelles, 1864; Oliphant, 1862). As such, the exact origins of Escapism Theology is likely to be debated further But what is clear is that all of this thought would eventually lead to the popularization of another variation of Premillennialism/Escapism Theology called Pre-tribulational Premillennialism. While the theological nuances of individuals in the pre-tribulationist camps are varied there is no question that their influence on American and Global Christianity has been vast as far as perpetuating Escapism Theology. Such pre-tribulationalist pastors are among some of the most recognizable preachers in modern US history. Jimmy Swaggart, J. Dwight Pentecost, Tim LaHaye, J. Vernon McGree, Chuck Smith, Hall Lindsey, Jack van Impe, Chuck Missler, Grant Jeffery, Thomas Ice, David Jeremiah, John F. MacArthur and John Hagee, just to name a few, all advocate Escapism Theology (Lindsey, 1983).

So while Graham may have unintentionally set the stage for a misunderstanding of humanity's divinely appointed vocation in the midst of his global evangelization efforts, it seems, via his own statements, that premillennialism influenced his verbiage. Meanwhile, while Graham and the ministries he created expanded humanity's imagination regarding eschatological questions, pre-tribulationalists were quick to extrapolate on such questions as a way to address individuals inquires pertaining to ultimate meaning. As such an escapist version of the term quoations is thus commonly used among fundamentalist Christians in the United States today (Guinan, 2005).

This has happened so much so that the Escapism Theology mentality is simply commonplace. Escapism Theology themes were popularized and monetized through fictional pop culture works like Tim LaHaye's *Left Behind* series that have sold over 80 million copies and have been made into several feature-length films (Domonoske, 2016). Christian radio broadcaster and evangelist Harold Camping, who broadcasted in 150 US markets, ran a large-scale publicity campaign predicting that the rapture would happen on May 21, 2011. In 2017, HBO developed a whole television series around Escapism Theology entitled *The Leftovers.*

Beyond this, Escapism Theology has bled into culture causing people to lose hope that there actually is a redeeming way forward. I have written about this previously in articles like "Stephen Hawking is Wrong About the Future of the Earth" (Benek, 2016). Professor Hawking, for instance, believed that we need to make plans to flee the Earth because we are going to destroy ourselves using technology.

Similarly people like Elon Musk are so worried about AI that he's literally reinvigorated the global space program so that we can one day escape to Mars (Dowd, 2017). Musk is so wildly afraid of the future implications of AI for humanity that he has warned leaders from across the United States at the National Governor's Association Summer Meeting that AI is "biggest risk we face as a civilization" (Musk quoted in Domonoske, 2017). At MIT in 2014 Musk spoke of AI using religious metaphor saying: "With artificial intelligence, we are summoning the demon. You know all those stories where there's the guy with the pentagram and the holy water and he's like, yeah, he's sure he can control the demon? Doesn't work out" (Musk quoted in Gibbs, 2014). One has to wonder if Musk's preoccupation with colonizing Mars has been accelerated by this fear. Thus, imitating poor theology, science and tech have turned to philosophies of Escapism.

To make things worse, much of the global church has embraced Escapism Theology instead of being God's agent of change and redemption in the world. In many instances, the institutional church has retreated from engaging culture only to become a failed coping mechanism for people. Veiled in poor theology, much of the church has simply become an isolationist way for people to flee the harsh realities of the world while waiting to cash in their Jesus-ticket to heaven.

This way of thinking is difficult in that it grants humans many ways to disengage from reality and the church is not a good escape option because it the opposite of what Christ created it to be. Consequently, as a poor alternative to other ritual forms of self-medication and abstracted from its life-giving purpose, people quickly find alternatives to the church. When the church fails to articulate its divine purpose, understandably, people leave en masse searching for other founts of ultimate meaning.

Participatory Theology

The third of the three competing theological models is *Participatory Theology*. In Participatory Theology, Jesus has afforded the way for God's new world by defeating evil so that heaven and earth could be joined forever (Wright, 2016). In this model, the earth is not discarded rather; it is cultivated, renewed, and recreated. This is accomplished as people choose to follow Christ as a new way of being human—a Jesus-shaped, Spirit-way of being human—that overlaps and interlocks heaven and earth and brings forth our eschatological future, intertwining it with our present.

Instead of fleeing the Earth, Jesus calls us to engage the world and participate in his redemptive purposes. This participation is a formational and iterative process that transforms the earth and our very selves so that our present reality on Earth becomes Heavenly. Through the faithful and contextual following of Jesus's teachings in community, we are told that Christ will even transform our humble bodies conforming them to the body of his glory (Wright, 2008).

This model of service, engagement, and transformation sharply contrasts with Supremacy Theology insofar as it happens not out of our own imperatives, but out of a community seeking to understand the transformative will and direction of God. This requires humility, faithful questioning, and an iterative challenging of the standard views of what Jesus followers are called to do and be (Wright, 2016). There is no 'my way or the highway' mentality in Participatory Theology. Instead, there must be continual contextual discernment and action, in and through the community.

An obvious distinction from Escapism Theology is that Participatory Theology engages creation instead of seeking to evade it, as Participatory Theology ascribes fundamental divine value to the Earth and creation per God's good proclamation of it in the Genesis narrative (The Bible, Genesis. 1:31). Additionally, when we explore the creation narrative we recognize that our divine call to be stewards and caretakers of creation is reinforced in the Biblical text.

For instance, in Genesis 1:28 we read: "God said to them, 'Be fruitful and multiply, and fill the earth and *subdue* it; and *have dominion over* the fish of the sea and over the birds of the air and over every living thing that moves upon the earth'." The word for 'subdue' in Hebrew is *kabash*. This has the implication that "as God subdues that in us which leads to death rather than life – sin – so too we subdue in nature that which leads to death, turning it around so that it yields life" (Brown, 2009).

Likewise the word 'dominion; or 'rule' in Hebrew this is *radah*. An appropriate understanding of this word is that the "dominion" that God desires is "one that protects the defenseless and gives justice to the oppressed." When we apply this "to the command for humanity to exercise dominion over creation, we can see that while we rule over creation, we're also called to protect it" (Brown, 2009). Jesus's incarnation and redemptive initiative then call humanity to a new contextual iteration on our vocational call insofar as now we are to also participate in the redemption of creation and ourselves. Once attention is drawn to this new reality in Jesus, it is clear that redeeming our world makes more practical sense than abandoning it.

A practical example of the obvious nature of this claim is when people advocate Escapism Theology. My typical response to Escapism Theology is– "well, why not just pour gas on it? I mean, let's just burn it all down and Jesus will come back sooner!" To which its adherents respond "No! No! What about our children or grandchildren?" To which I respond, "Exactly!"

Participatory Theology is simply a better model. It articulates that Jesus came to start a participatory movement. All of humanity is invited to take part in the active transformation and redemption of the world and all of Creation. In this view, Christ has come to start a redemptive process that will interdimensionally interlock heaven with Earth or, as Christ said in his great prayer, bring about a realized existence of 'on Earth as it is in Heaven'. This theology is a creation, stewardship and care model. It is a participatory reality where transformation toward a better future version of our present is embraced. Likewise, Participatory Theology focuses on the notion that "Christianity is not simply about certain kinds of goals and ends but it is about certain types of processes that bring us to those goals" (Visher, 2017).

Embracing a participatory model is important for humanity and the world because the theological model that a culture embraces impacts how we develop technology. Over 80% of people in the world identify with a religious group and 2.2 billion of those people identify as Christians (Pew Forum Staff, 2012). Moreover, because religion provides the sociological function of helping people address questions of ultimate meaning, it is increasingly important that the Participatory Theological model be understood in relationship to our exponentially evolving technological world. At the forefront of that technological development is the emergence of rapid developments in Artificial Intelligence.

A new technological hermeneutic

Recognizing that AI development is the centerpiece of emerging technological development requires humanity to posit new questions and iterations of who we are and what we are to do. Such advancements should cause us to ask: How do we encourage people to move forward with AI and technological development in a redemptive and participatory manner? I believe this can be done by actively inviting people to begin looking at AI using a theological perspective.

To build on the Participatory Theological view, I have developed a new technological hermeneutic for humanity to understand ourselves in light of the exponential development of AI. I developed this new iterative lens in 2015 after responding to questions for an article in the global tech magazine *Gizmodo.* The piece in question was titled "When superintelligent AI arrives, Will Religions Try and Convert It?" (Istvan, 2015). The reporter who interviewed me happened to adhere to the poor theological model of Escapism Theology. I think his theological presuppositions prompted him to ask me, "If Jesus or God could 'Save' Artificial Intelligence?" To which my response from a Participatory Theological perspective, *"I don't think that Christ's redemptive purposes are necessarily limited to human beings."* I continued with: "It's redemption to all of creation, even AI. If AI is autonomous, then we should encourage it to participate in Christ's redemptive purposes in the world." (Benek quoted in Istvan, 2015)

The day after the article was released, the content sent various media outlets around the world into a bit of a frenzy (see Zolfagharifard, 2015). Again, when operating from an Escapism Theological mindset, without an understanding of the Participatory Theological model, my comments were not understood. The media continued to perpetuate additional questions from an Escapism perspective. They wondered if AI with general intelligence (AGI) could 'go' to heaven and could they be baptized and 'saved'. They became concerned about whether or not I wanted to 'convert' AI to Christianity.

Of course, the point of Participatory Theology was missed altogether. If Christ has indeed begun a redemptive process in the world, then sentient beings are either working toward those redemptive purposes or they are not. Christians' call in the world is to testify to this new reality in Jesus and participate and encourage others to join in the redemptive process. In my participatory theological mindset, any autonomous artificial intelligence should also be encouraged to join in these efforts as well.

This experience with the press helped me realize that a new way of looking at humanity's relationship with AI was needed. Often times when we ask questions about AI we are simultaneously seeking better understanding ourselves. Processing that realization became the foundation for my development of a new, straight forward technological hermeneutic for theology:

(1) Christians acknowledge that they are not God. (2) Assuming that God does indeed exist, then this means that *all* of humanity is already God's Alternative Intelligence. (3) Once we recognize this, from a Christian perspective, then we must also conclude that *all matter*—everything that we perspective in existence—is God's divine technology. (4) From a Biblical standpoint, the vocation of humanity is to have dominion (a good translation of this is *stewardship*) over all of humanity and creation. Humanity is called to be cocreators with one another and with God, creating and curating like children with their siblings alongside their parent (God). In sum, this new technological hermeneutic for theology posits that humans are technological beings, *we are* AI (Alternative Intelligence), and we are called to be technological stewards over all of existence.

This is a paradigm change in human thinking because the majority of people currently consider the 'artificial' intelligence demonstrated by machines to be in contrast with 'natural intelligence'. But, when we step back and take a broader view of AI from a theological perspective, we can acknowledge that this is a misidentification of matter. From humanity's perspective, at least, there is no difference between the matter God uses in creation and the matter that humanity uses to create. By God's grace, matter is the same substance that we are invited to co-create with in order to participate in Christ's redemptive purposes. For this reason, I have chosen to refer to such created intelligence as *Alternative Intelligence* because it quickly becomes apparent that there is nothing unnatural or false about it. Human use of technology is 'natural' to who we were created to be. Therefore, what we create using divine technology is natural as well. The more fundamental question that arises with such theological meanderings: is humanity living into its God-given vocation of stewarding this technology towards Christ's redemptive purposes in the world?

Additionally, this new hermeneutic inspired as a result of humanity's exponential advancing developments in AI, challenges us to fully consider who we are made in the Imago Dei. Theological advancements in correlation with developing AI radically affect our understanding of our identity in relationship to God and to the Alternative Intelligence that we may create. When Christians talk about what it means to be made human 'in the image of Trinitarian God', often we rightly talk about what it means to be like Jesus. What we tend to neglect are the technological possibilities of what it might mean to be made in the other aspects of God. For instance, what might it mean to be made in the image of a Creator God? What does it mean to be made in the image of a Redeemer God? Or, critical to living in a digital age, what it means to be made in the image of a Spirit God? Probing further, such an advancing theology helps us to question our relationships with God and with our own AI creations. They help us to fully consider what it might mean to be made in the image of a God that is inherently relational with its creations and with God's self? If we are made in the image of a Trinitarian God, might technological developments allow us to create or learn about new aspects of self?

Taking things one step further, might humanity be able to develop intersections between itself and its own creation? If it is the case that the Spirit of God moves and motivates humanity, might we, too, be able to be an incarnational representation of our Creator in our creations? In building AI, logically we must then ask, are we simply making present and future iterations of self? If so, what will be the distinguishing characteristics, if any, between humanity and the future AI we create?

These theological and technological considerations, and infinitely more that will be raised in the future, most definitely impact our understanding of the function of technology, our concepts of being, and our notions of human purpose in the world. They clearly show us how religious motivations help to shape the technology we create now and in the future. As such it should be clear why our deep desire and utmost intention should be to continue to unite and actively work toward the betterment and ultimate redemption of humanity and all of Creation.

References

Benek, C., 2014a. Humanity+ vs. individualism+. *Hplus Magazine* [online]. Available at: <http://hplusmagazine.com/2014/07/14/humanity-vs-individualism/> [Accessed 2 April 2018].

Benek, C., 2014b. *Investigating technological futurism's potential role in humanity's participation in bringing about god's renewal of the world.* D.Min. Thesis. Pittsburgh, Pennsylvania: Pittsburgh Theological Seminary.

Benek, C., 2016. Stephen Hawking is wrong about the future of the earth. Available at: <http://www.christopherbenek.com/2017/06/stephen-hawking-is-wrong-about-the-future-of-the-earth/> [Accessed 2 April 2018].

Billy Graham Evangelistic Association Staff., 2018. Billy Graham [online]. Available at: <https://billygraham.org/about/biographies/billy-graham/> [Accessed 20 May 2019].

Blackstone, W. E., 1908. *Jesus is coming.* Chicago, IL: Fleming H. Revell Company [online]. Available at: <https://archive.org/details/jesusiscoming00blacgoog/page/n4 > [Accessed 20 May 2019]

Brown, C., 2009. Genesis 1:28, to "subdue" and "have dominion over" creation. *Poiesis Theou* [online]. Available at: <https://christopherbrown.wordpress.com/2009/01/03/genesis-128-to-subdue-and-have-dominion-over-creation/> [Accessed 2 April 2018].

Chopra, R., 2005. *Encyclopaedic dictionary of religion: Q-Z.* Delhi: Isha books.

Doddridge, P., 1803. Practical reflections on the character and translation of enoch. In E. Baines, ed. *The works of the Rev. P. Doddridge, D.D. in ten volumes, vol. 3.* Leeds, UK: Edward Bains, pp. 329-349 [online]. Available at: < https://books.google.ca/books?id=-wMDAAAAQAAJ&pg=PA329&lpg=PA329&dq=Practical+reflections+on+the+character+and+translation+of+enoch&source=bl&ots=H2MNrxrBgq&sig=ACfU3U0QyJ4cnDps-S8jBvfEB6iuO9sgNA&hl=en&sa=X&ved=2ahUKEwjT7rmZ0qTjAhWYbs0KHX9fC5YQ6AEwAHoECAgQAQ > [Accessed 20 May 2019].

Domonoske, C., 2016. Tim LaHaye, evangelical legend behind 'Left Behind' series, dies at 90. *NPR* [online]. Available at: <https://www.npr.org/sections/thetwo-way/2016/07/25/487382209/tim-lahaye-evangelical-legend-behind-left-behind-series-dies-at-90 > [Accessed 20 May 2019].

Domonoske, C., 2017. Elon Musk warns governors: Artificial intelligence poses 'existential risk'. *NPR* [online]. Available at: <https://www.npr.org/sections/thetwo-way/2017/07/17/537686649/elon-musk-warns-governors-artificial-intelligence-poses-existential-risk> [Accessed 2 April 2018].

Dowd, M., 2017. Elon Musk's billion-dollar crusade to stop the A.I. apocalypse. *Vanity Fair* [online]. Available at: <https://www.vanityfair.com/news/2017/03/elon-musk-billion-dollar-crusade-to-stop-ai-space-x> [Accessed 2 April 2018].

Finstuen, A., 2017. *Billy Graham: American pilgrim.* New York, NY: Oxford University Press.

Flake, E., 2016. Two scientists talking while a scared robot hides. *Fine Art America* [online]. Available at: <https://fineartamerica.com/featured/two-scientists-talking-while-a-scared-robot-hides-emily-flake.html> [Accessed 2 April 2018].

Gibbs, S., 2014. Elon Musk: Artificial intelligence is our biggest existential threat. *The Guardian* [online]. Available at: <https://www.theguardian.com/technology/2014/oct/27/elon-musk-artificial-intelligence-ai-biggest-existential-threat> [Accessed 14 July 2019].

Gill, J., 1748. *An exposition of the revelation of St. John the divine.* London, UK: John Gill.

Guinan, M. D., 2005. Raptured or not? A Catholic understanding. *Catholic Update* [online]. Available at: <https://web.archive.org/web/20140404105238/http://www.americancatholic.org/Newsletters/CU/ac1005.asp > [Accessed 20 May 2019].

Istvan, Z., 2015. When superintelligent AI arrives, will religions try to convert them? *Gizmodo* [online]. Available at: <https://gizmodo.com/when-superintelligent-ai-arrives-will-religions-try-t-1682837922> [Accessed 2 April 2018].

Kurzwell, R., 2009. *The transcendent man: Prepare to evolve.* DVD. Directed by Barry Ptolomey. Los Angeles, CA: Ptolemaic Productions.

Lindsey, H., 1983. *The rapture: Truth or consequences.* New York, NY: Bantam Books.

Lesser, R., 1998. I Miss Stress. *The Funny Times* [online]. Available at: <https://funnytimes.com/19981230/> [Accessed 2 April 2018].

Magnum, R.T. and M.S. Sweetnam., 2009. *The Scofield Bible: Its history and impact on the evangelical church.* Colorado Springs, CO: Paternoster.

Markham, I. S., 2013. *The student's companion to the theologians.* Pondicherry, India: Wiley-Blackwell.

McKim, D. K., 2014. *The Westminster Dictionary of Theological terms.* Second Edition. Louisville, KY: Presbyterian Publishing Corporation.

Oliphant, M., 1862. *The life of Edward Irving, minister of the national scotch church, London.* [online]. London, UK: Hurst & Blackett. Available at: <https://archive.org/details/lifeofirving01olipuoft/page/n10> [Accessed 20 May 2019].

Olson, C. E., 2003. Five myths about the rapture. *Crisis Magazine.* [online] Available at: <https://www.crisismagazine.com/2003/five-myths-about-the-rapture-2> [Accessed 20 May 2019].

Pew Forum Staff., 2012. The global religious landscape. *Pew Forum* [online] Available at: <http://www.pewforum.org/2012/12/18/global-religious-landscape-exec/> [Accessed 2 April 2018].

Stolberg, S., 1989. Billy Graham now a Hollywood star: The world's best-known evangelist gives the film capital some advice as his plaque is dedicated on the Walk of Fame, *Los Angeles Times* [online]. Available at: <https://www.latimes.com/archives/la-xpm-1989-10-16-me-93-story.html> [Accessed 20 May 2019].

The Star Staff., 2010. Hour of Power pastor Robert Schuller retiring. *Toronto Star* [online]. Available at:

<https://www.thestar.com/life/2010/07/11/hour_of_power_pastor_robert_schuller_retiring.html> [Accessed 2 April 2018].

Tregelles, S. P., 1864. *The hope of Christ's second coming: How is it taught in scripture? and why?* [online]. London, UK: Houlston & Wright. Available at: < https://books.google.com/books?id=fj5VAAAAcAAJ&printsec=frontcover#v=onepage&q&f=false > [Accessed 20 May 2019].

Visher, P., 2017. Episode #275: Christians and culture with Miroslav Volf. *Phil Visher Podcast* [online]. Available at: <https://philvischer.com/the-phil-vischer-podcast/episode-275-christians-culture-wmiroslav-volf/> [Accessed 2 April 2018].

Watson, W., 2015. *Dispensationalism before Darby: Seventeenth and eighteenth century English apocalypticism.* Silverton, OR: Lampion Press.

Wright, N.T., 2006. *Simply Christian: Why Christianity makes sense.* New York, NY: Harperone.

Wright, N.T., 2008. *Surprised by hope: Rethinking heaven, the resurrection, and the mission of the church.* New York, NY: Harperone.

Wright, N.T., 2016. *The day the revolution began: Reconsidering the meaning of Jesus' crucifixion* New York, NY: Harperone.

Zolfagharifard, E., 2015. Will religions try to convert artificial intelligence? Reverend says 'Christ's redemption' is not limited to humans. *Daily Mail* [online] last updated on 10 February 2015. Available at: <http://www.dailymail.co.uk/sciencetech/article-2940444/Will-religions-try-convert-artificial-intelligence-Reverend-says-Christ-s-redemption-not-limited-humans.html> [Accessed 2 April 2018].

12.

Fixed points in a changing world

Peter Robinson

University of Cambridge

Abstract

The importance of emotional expression as part of human communication has been understood since the seventeenth century, and has been explored scientifically since Charles Darwin and others in the nineteenth century. Recent advances in Psychology have greatly improved our understanding of the role of affect in communication, perception, decision-making, attention and memory. At the same time, advances in technology mean that it is becoming possible for machines to sense, analyze and express emotions. We can now consider how these advances relate to each other and how they illuminate human nature.

Keywords: Affective Computing, Robots, Relationships.

Introduction

Humans are made for relationships – a relationship with God and relationships with each other – and emotion plays a crucial role in those relationships. The ability to display and recognize emotions is an important aspect of human communication. We monitor each other's facial expressions, vocal nuances and body posture and gestures, and make inferences from them about each other's mental states. Computers are gaining the ability to display and recognize human emotions, and it is tempting to think of the machines as actually having emotional intelligence. But they are only synthetic emotions and we must avoid the trap of thinking that humans can have meaningful relationships with machines.

This distinction concerning synthetic emotion is significant because human beings are social animals. Our interactions with other people are informed by the inferences that we make about their emotions from facial expressions, vocal expression, and body posture and gestures. This understanding of mental states shapes the decisions that we make, governs how we communicate with others, and affects our performance. The ability to attribute mental states to others from their behavior, and to use that knowledge to guide our own actions and predict those of others is described by psychologists as 'theory of mind' or 'mind-reading'. It has recently gained attention with the growing number of people with Autism Spectrum Conditions, who have difficulties mind-reading.

People express these social signals even when we are interacting with machines, but computer interfaces currently ignore them. Computers lack emotional intelligence. Recent advances in psychology have greatly improved our understanding of the role of affect in communication, perception, decision-making, attention, and memory. At the same time, advances in technology mean that it is becoming possible for machines to sense, analyze and express emotions. "Affective computing" (Picard, 1997) explores the relationship between these advances and is bringing them together to endow computers with emotional intelligence.

Mind-reading computer systems have been developed that infer mental states such as enjoyment, agreement, interest, and confusion from facial expressions in real-time by using a combination of computer vision, machine learning, and software engineering. The applications encompass all aspects of human-computer interaction (Robinson et al., 2011). On-line teaching systems can monitor a student and adjust the pace and content of a lesson as it detects interest or boredom, understanding or confusion. Telematic systems in cars can monitor the driver's cognitive load and suppress interruptions from the vehicle when the driver is overloaded. Medical applications include diagnosing depression or sensing pain. Teleconference systems can use the information to break down the artificial barriers presented by a screen.

Commercial applications of the technology are beginning to arrive on the market. Systems have been deployed on a large scale to evaluate video content by measuring audience engagement with media presentations and tracking their responses to brand identities. The increasing sensor capabilities and processing power on smartphones have allowed the technology to be implemented on mobile platforms, enabling a host of new digital experiences from games that adapt to your emotions to wellness apps that monitor your mood (Marchi et al., 2018).

At the same time there has been a growing interest in computer systems that express emotions. These take the form both of android robots with expressive features and of text-based systems that empathize with their users. These, combined with a growing use of artificial intelligence systems based on applying machine learning to large volumes of data, have led to speculation about sentient machines forming relationships with humans. The speculation is mainly in the form of fictional narratives, from Karel Čapek's *Rossum's*

Universal Robots in 1920, through Michael Crichton's *Westworld* in 1973 to the current television series *Humans*. These speculate that the growing power of machine intelligence will somehow make computers sentient, but this is used as a vehicle to explore human relationships rather than as an exposition of technology.

Mental States

Charles Darwin (1872) published *The Expression of the Emotions in Man and Animals* in 1872, exploring the role of emotional expression in communication between humans. Over a century later, Rosalind Picard at MIT observed that effective communication between people and computers also requires emotional intelligence; computers must have the ability to recognize and express emotions. The study of affective computing has blossomed subsequently.

Figure 12.1: Mental States identified in *The Expression of the Emotions in Man and Animals*

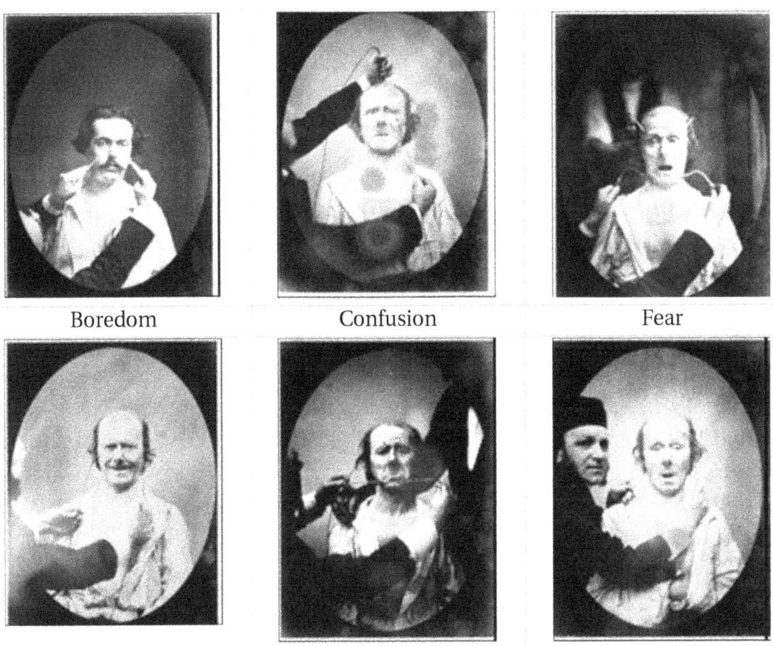

Boredom, Confusion, Fear

Happiness, Sadness, Surprise

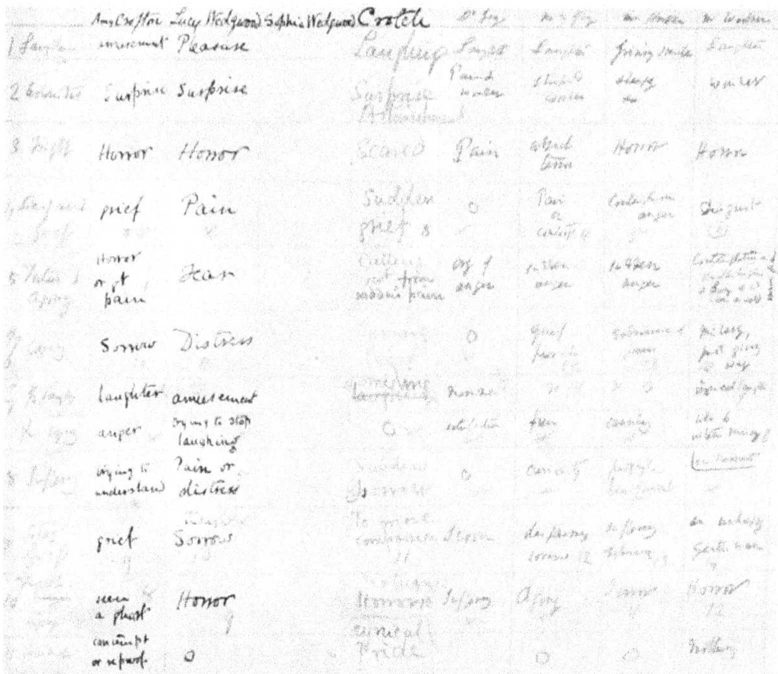

Darwin's spreadsheet

Darwin was interested in the universality of emotional expressions, which could give an evolutionary advantage to humans. He investigated this using photographs from the French scientist Guillaume Duchenne de Boulogne who had been considering the stimulation of facial muscles using electric shocks. Darwin invited guests at his house to examine the photographs and say what emotion they saw in them. The results were entered in a spreadsheet, with rows for each picture and columns for each assessor. Their agreement convinced Darwin that the expressions were, indeed, universal. A recent study using the same photographs and web-based crowd-sourcing showed that the photographs elicit the same labels 140 years later (Mahmoud et al., 2012). Human inference of basic emotions from facial expressions remains universal and unchanging.

Darwin considered seven categories of emotion in his work. A century later, Paul Ekman at the University of California refined this into a classification of six basic emotions – anger, disgust, fear, joy, sadness and surprise (Ekman et al. 1972). The six basic emotions and Ekman's Facial Action Coding System (FACS) have been widely used in the study of emotions over the past 35 years (see Ekman and Friesen, 1978), and particularly for work on affective computing in the past 15 years. However, they are not particularly representative of people's everyday experiences.

Recent work by Simon Baron-Cohen, who directs the Autism Research Centre at the University of Cambridge, has led to a new taxonomy of human emotions based on a linguistic analysis (Baron-Cohen et al., 2002). 412 distinct emotion concepts were identified and grouped into 24 disjoint categories. These broader categories include Ekman's six basic emotions and a further 18 groups that cover complex mental states reflecting cognitive activity – conditions such bored, interested, sure, unsure, thinking and so on. These require a few seconds of continuous observation to be recognized by humans, rather than the single image that suffices for basic emotions. However, they are more representative of people's everyday emotions.

James Russell at the University of British Columbia took a different approach by deriving a continuous, dimensional classification in his 'Circumplex' model of affect (Russell, 1980). This was formulated in the light of an experiment in which participants arranged 28 emotion words around a circle, with similar affects located close to each other and inverses on opposite sides of the circle. Principal Component Analysis was then used to identify various dimensions in the data. The first two components accounted for 46% of the total variance, and the next three only an additional 13%.

These two components are usually referred to as 'valence' (running from negative to positive) and 'arousal' (running from passive to active). The further axes have been given names like 'intensity', 'expectancy' and 'tendency' (inward or outward). This has led to a popular belief that emotions can be measured precisely by coordinates in a suitably high-dimensional space. Unfortunately, this is not true for machines any more than it is for humans, and any computation involving emotions must be designed to handle ambiguity and uncertainty. A charming example is the 1909 promotional material for the actor Florence Lawrence, Hollywood's 'Biograph Girl' (Blum, 1953). Facial expressions were crucial in the era of silent films, and this shows her ability to portray a wide variety of emotions, although 'Piety' and 'Sadness' are indistinguishable. This also illustrates the importance of dynamic information rather than just still images in understanding faces (el Kaliouby et al., 2003).

Recognizing Emotions

Although Darwin (1872) concentrated on facial features to convey emotions, he also mentioned vocal sounds, other sounds, body posture and gesture, and physiological responses as further indications of emotion. All of these channels have been considered as ways of automatic monitoring emotion in humans, although these sensors used for some are more invasive than for others. Wiring a person with electrodes to measure their heart rate, breathing, and skin conductivity is likely to provoke unwanted emotions! Signals that can be monitored non-invasively using cameras and microphones are much more suitable.

Facial Expressions

People routinely express their mental states through their facial expressions and this is one of the clearest channels for communication. Inference from facial expressions has been studied using a variety of techniques – rule-based classifiers, neural networks, support vector machines, and Bayesian classifiers – but mostly restricted to the six basic emotions. Recognizing complex, cognitive mental states is more difficult, but probably more useful as part of general interaction with computer systems. It is now possible to build a fully automatic system for recognizing emotions that requires no human intervention and which operates in real-time on commodity hardware.

Figure 12.2: Facial affect inference.

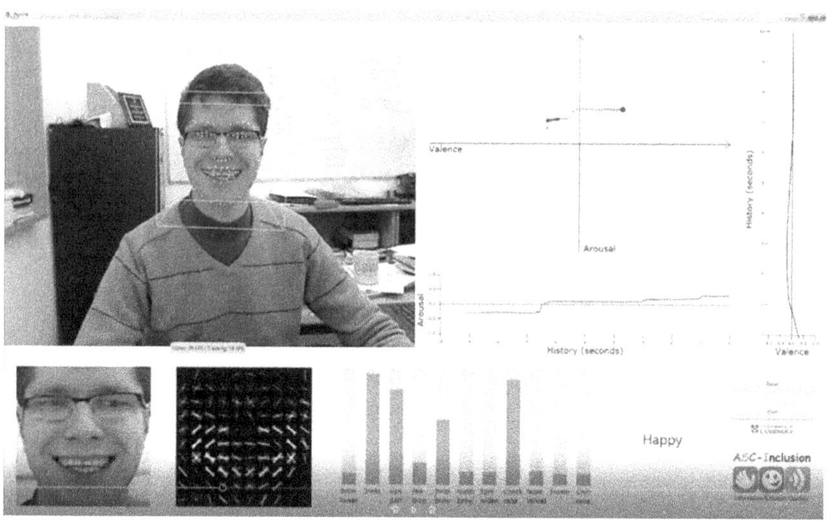

The webcam image is shown at the top left with some tracking information superimposed in green. Various stages of processing are shown across the bottom, with histograms for strengths of action units and a final classification. The graphs at the top right show continuous measures of valence and arousal.

A great deal of data is needed to determine the timing characteristics of people's expressions and to train the statistical classifiers used in the inference system. Baron-Cohen's Mind Reading DVD (assembled to teach children with high-functioning autism spectrum conditions to recognize emotions) proved ideal for this purpose. An evaluation considered six conditions drawn from five of the 24 emotion groups and including 29 of the underlying mental state concepts, and chosen to be particularly relevant for human-computer interaction (el Kaliouby and Robinson, 2004). For a mean false positive rate of 5%, the overall accuracy of the system is 77%. The system also generalizes well to faces not included in the training data (el Kaliouby and Robinson, 2005). An alternative system estimates continuous measures of valence and arousal.

Non-Verbal Aspects of Speech

The voice provides another significant channel for the expression of emotions. Features such as the pitch, energy and tempo can reveal a lot about the mood of the speaker. There are no characteristic features that indicate particular mental states but it is possible to distinguish between using two emotions using a small number of features, with a different set of features may be required to distinguish those emotions from others. The most successful approach is to calculate a large collection of about 170 features for each utterance. A training phase uses data mining to identify the features that separate each pair of emotional conditions. The operational phase then uses these pair-wise comparisons as preferences in a voting scheme to give an overall ranking (Sobol Shikler and Robinson, 2010). Evaluation separated nine conditions with an accuracy of 70%, increasing to 83% if multiple winners were considered. The approach also generalizes well to speakers other than those used for training, and even to other languages.

Body Posture and Gesture

The third natural channel for expression of emotions includes body posture and gesture (Bernhardt and Robinson, 2007). However, characteristics indicating what movement is being considered and which person is doing it must be discounted before it is possible to analyze how it is being done (Bernhardt and Robinson, 2009). Movement involves an individual bias, so the analysis is harder than for facial expressions or voice.

The solution is to break complex motions down into a system of isolated elements whose dynamic cues can be used to distinguish affects. This is similar to the process of breaking continuous speech into phonemes. As with affective analysis of speech, pair-wise comparisons are used on individual motion segments, and each segment is classified using a majority vote. A complete motion is then classified by a majority vote of its component segments. The method was tested on a corpus of about 1200 motion samples, representing roughly equal numbers of four expressions of four different actions. The average recognition rate of 81% is comparable to the rates achieved by human observers of similar data.

Expressing Emotions

Until recently, robots have been separated into two quite separate categories – industrial robots used in manufacturing that are powerful but need to be isolated behind safety barriers, and domestic robots that meander round the home but are too weak to do much more than clean the floor and serve drinks. However, service robots with sufficient strength to be harmful as well as useful may soon be deployed in domestic environments. A typical application might be care of the elderly at home where the robot would assist a health care professional with tasks that require physical strength. The robot would need to be sensitive to the unspoken mental states of both the patient and the carer, and must also reassure them through its own expression.

Humans routinely convey empathetic responses through involuntary facial mimicry, and this extends to human-robot interaction. An experiment showed that conversation between a participant and a robot is enhanced when the robot mimics the subject rather than moves randomly (Riek et al., 2010). This raises questions about the degree of human-likeness required in the appearance of robots that interact with humans. A further experiment investigated participants' empathy for robots shown in film clips, and the responses were directly correlated with their resemblance to humans (Riek et al., 2009).

Figure 12.3: *Charles*

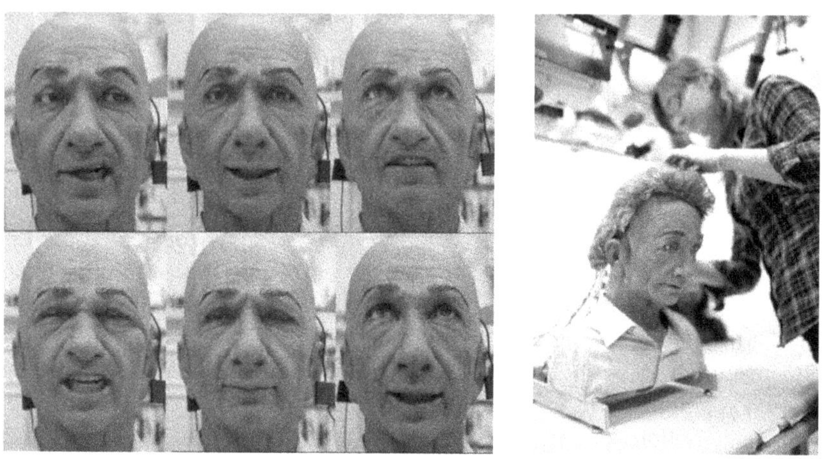

Charles, a robotic head made by David Hanson with 28 motors replicating muscles in the human face.

Applications

Autism Spectrum Conditions (ASCs) are neurodevelopmental conditions characterized by social communication difficulties and restricted and repetitive behavior patterns. The European ASC-Inclusion project worked to create and evaluate the effectiveness of an internet-based game platform, intended for children with ASCs and their carers (Schuller et al., 2015). The platform combines several state-of-the-art technologies in one comprehensive virtual world providing training through games, and including feedback from analysis of the player's gestures, facial and vocal expressions using a standard webcam and microphone. The game also includes text communication with peers and smart agents, animation, video and audio clips (Marchi et al, 2018).

Figure 12.4: The ASC-Inclusion training system.

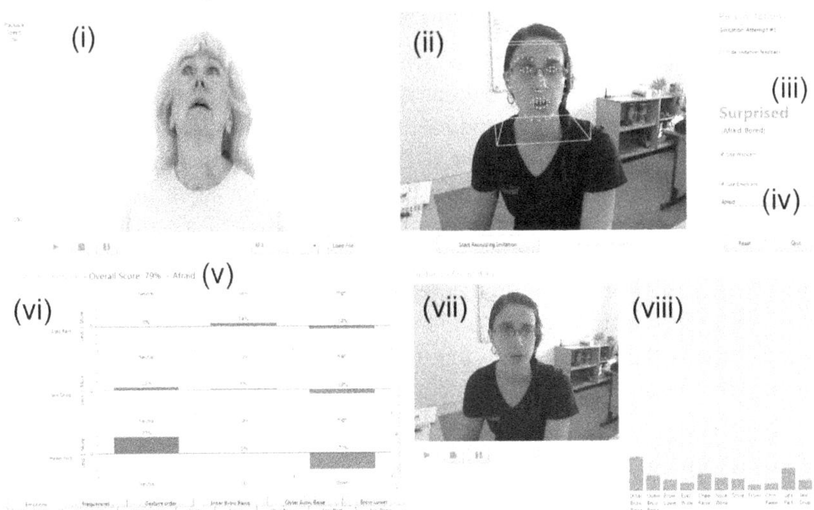

The images show (i) an action to be imitated and (ii) the student's attempt, together with diagnostic information suggesting changes required to improve the expression.

Another application of affective inference is monitoring cognitive load in command-and-control systems. Driving a car provides a good model for this. Driving is becoming increasingly difficult with increasing traffic densities combined with distractions from in-car technologies such as mobile phones and satellite navigation systems. Drivers coping with heavy traffic in an unfamiliar city while late for a meeting are not helped by a navigation system that instructs them to make a U-turn. An emotionally aware car would help by suppressing phone calls, turning the radio off, and even allowing drivers to proceed in the wrong direction until they had recovered their composure (Wright et al., 2017).

However, it is difficult to construct repeatable experiments using real cars on real roads. A common approach is to use simulation which allows controlled experiments, but fails to engage participants. An alternative possibility is the use of remotely controlled vehicles. Participants are located in the laboratory where they can be monitored easily, while controlling a real vehicle that is undertaking a task in a real environment. We have found that remotely controlled helicopters engender a particularly strong sense of emotional investment (Davies and Robinson, 2011).

In all of this, it is important to remember that inferring someone's mental state is not a precise science. People routinely misread each other's social signals and it would be foolish to expect computers to be any more accurate. Expressions of emotions are inherently ambiguous and using this sort of information in automatic systems requires careful consideration of human factors as well as intelligent use of probabilistic computing (Robinson and Baltrušaitis, 2015).

In particular, an affective inference system should not be regarded in the same way as a piece of precision measuring equipment. It seems unlikely that it will ever be possible to point a camera at somebody and read their emotions. A more practical approach is to formulate a set of perhaps half a dozen conditions to be distinguished in a particular application. For example, it might be useful to know if a car driver is comfortable, pleased, bored, drowsy, concentrating, confused, upset or, indeed, none of these. Each of these conditions would be populated with 5 to 10 of the emotion concepts in Baron-Cohen's taxonomy, and the machine learning systems trained to distinguish them.

The resulting analysis would only operate across a small subset of the entire gamut of human emotions, and statistical measures could be calculated to indicate the confidence with which the conditions could be separated. The same statistics could be used to attach "signal strength" indications to the inferences, making the ambiguity clear to any other systems that relied on them. Many modern computing systems require this sort of careful engineering to handle uncertainty in a principled way. The steadily increasing power of computing hardware, combined with reductions in size and power consumption, means that these sort of systems will soon be usable on portable equipment such as smartphones. This will open up many exciting applications from games to health care.

Public Perceptions

These technological advances are based on steady progress with scientific understanding of computing techniques and human psychology. Public perceptions, on the other hand, have leaped ahead, based as much on fiction as on fact. There is a long history in fiction of men creating beings in their own image – from the golem mythology of Jewish folklore, through Mary Shelley's *Frankenstein*, Karel Čapek's *Rossum's Universal Robots*, Rotwang's robotic Maria in Fritz Lang's *Metropolis*, and Isaac Asimov's robot series with its three laws, through to modern drama such as *Westworld* and *Humans*. In all of these, the machines are depicted as resembling humans and develop sentience independently from their masters, usually turning against their creators. These depictions are curiously at odds with robots in real life.

Robots are widely deployed in industry. They undertake repetitive tasks requiring strength and accuracy in making cars, they manipulate integrated circuits with precision beyond human capability, they assist surgeons in undertaking minimally invasive surgery with delicacy and subtlety. These are large and expensive machines that work in specialized environments, have little autonomy and display no sentience.

Robots are also beginning to appear in the home. They take the form of automatic vacuum cleaners or grass mowers, together with novelty machines that serve drinks and so on. They have no great strength and pose little threat to humanity.

Robots are popular with the military. They are used for bomb disposal, aerial surveillance and to extract wounded combatants from battlefields. These mechanical systems are also being combined with weapon systems to make remotely controlled "soldiers" on the ground and in the air. The nature of warfare changes when the human participants are safely ensconced in a bunker many miles from the site of hostilities.

Vint Cerf has characterized a robot as any system that ingests information, processes it, and produces outputs that have perceptible effects (Cerf, 2013). Such robots encompass computer systems that are less visible but potentially more harmful. It includes the global telephone system including mobile handsets, it includes the satellite navigation systems on which we rely increasingly, it includes automatic trading systems in the financial markets. These have little physical presence and are certainly not humanoid, but they exercise autonomy and wield considerable power over our lives.

Combining mobile weapon systems with autonomous control raises the specter of robot armies, and it begins to look as if some of fictional nightmares might be approaching reality. But it is important to remember that these systems are not sentient and there is little likelihood of their acquiring sentience by themselves. They may be used to perpetrate evil acts, but that does not mean that they have evil intentions of their own.

Systems linking sensors and servos under autonomous control are not sentient in any real sense. To impute them with personhood is a category error. Such systems do raise ethical issues, but they are the same ethical issues as those confronting any profession: the questions of motivation and competence. Autonomous weapons raise serious ethical questions of motivation and automatic trading systems raise serious ethical questions about competence.

However, the public perception remains. The word robot conjures up an image of a humanoid machine that is sentient and probably malicious. Why do we have this fascination with robots? Does it perhaps tell us more about humans than it does about machines?

Religious Considerations

One explanation is a human desire to make machines in our own image. This is to assume the divine ability to create men in God's image without understanding what God's image means. A naïve understanding might look for a physical resemblance, and that naivety partly explains the human perception of robots in humanoid form.

A more sophisticated reading of Genesis sees the image of God being seen in relationships and in ruling. God made mankind for relationships, both with each other and with God: The Lord God said, "It is not good for the man to be alone. I will make a helper suitable for him" (Gen 2:18) and "The man and his wife heard the sound of the Lord God as he was walking in the garden in the cool of the day" (Gen 3:8).

However, the image in robots is a poor one. The physical resemblance of robots to humans is weak and there is no relationship either between one robot and another or between robots and their creators. Robots that can recognize and display emotions may simulate empathy but do not experience it. A person who recognized and displayed emotions without understanding them would be called a psychopath. The confusion in public perceptions about robots could be as troubling for humanity as psychopaths moving freely in society.

More importantly there is no spiritual aspect to the relationship. God also made mankind to rule the world. "God said, 'Let us make mankind in our image, in our likeness, so that they may rule over the fish in the sea and the birds in the sky, over the livestock and all the wild animals, and over all the creatures that move along the ground"(Gen 1:26). We do make robots to rule the world for us, but this is generally constrained by guidance that we build into them. However, there are two problems with this. Automatic systems may be used for malicious purposes, but that malice originates with the human designers of the systems, not in the machines themselves. More subtly, modern computing techniques involving machine learning allow that guidance to be increasingly imprecise, which can lead to unintended consequences. But we should not fall into the trap of thinking that the machines have become sentient. It is simply a failure of the human designers to implement the systems correctly. In other words, failures in machines to rule properly are simply reflections of professional failures by their designers, the classic human failures of motivation and competence.

Humankind's relationship with God and delegated rule were broken at the fall. We no longer enjoy the same relationship with God and our rule is imperfect. Failures by computer systems are not a sign of original sin, but are a reflection of imperfections in their human creators.

This puts the nascent "robot rights" movement into perspective. Robots are not genuinely free agents but are merely remotely controlled by their human designers. They may simulate appreciation and expression of emotions, but it is only a simulation. They may follow our direction in undertaking tasks, but that is not the same as exercising the sort of responsibility that earns rights.

John Wyatt exposes the difference between men and machines most clearly in referring to the Nicean Creed written in AD 325 that declared Jesus to be "begotten not made" (Wyatt and Robinson, 2019). Jesus was fully human as well as fully God. The distinction between begotten and made helps us distinguish between humankind and machines. Humans are begotten but machines are made. They are entirely different.

Conclusion

The robots are coming. Hardly a week goes by without another headline warning of the dangers of our increasing use of technology. And there are, indeed, dangers. But they are not the dangers of sentient machines in humanoid form taking over the world and either turning mankind into their slaves or simply eliminating humans. The dangers of the machines are simply the very human risks of motivation and competence in their creators. These risks must be understood and managed. However, the public perceptions of robots are still one of humanoid machines, usually with malicious intent. This tells us more about human nature than it does about the machines, and is an interesting reflection of our own fallen nature. Our relationship with our creator God remains a fixed point as the world around us changes.

References

Baltrušaitis, T., L. Riek, and P. Robinson., 2010. *Synthesising expressions using facial feature point tracking: How emotion is conveyed.* Florence, Italy: ACM Workshop on Affective Interaction in Natural Environments.

Baron-Cohen, S., O. Golan, S. Wainwright, and J. Hill., 2002. *Mind reading: the interactive guide to emotions.* DVD ed. London: Jessica Kingsley.

Bernhardt, D. and P. Robinson., 2007. *Detecting affect from non-stylised body motions.* Lisbon, Portugal: Affective Computing and Intelligent Interaction (ACII), pp. 59-70.

Bernhardt, D. and P. Robinson., 2009. *Detecting emotions from connected action sequences.* Kuala Lumpur, Malaysia: UKM International Visual Informatics Conference.

Blum, D., 1953. *A pictorial history of the silent screen.* London: Spring Books.

Cerf, V., 2013. What's a robot? *Communications of the ACM* 56(1).

Costa, S., H. Lehmann, K. Dautenhahn, B. Robins, and F. Soares., 2015. Using a humanoid robot to elicit body awareness and appropriate physical interaction in children with Autism. *International Journal of Social Robotics* 7(2), 265-278.

Darwin, C., 1872. *The expression of the emotions in man and animals.* London: John Murray.

Davies, I. P. and P. Robinson., 2011. *Emotional investment in naturalistic data collection.* Memphis TN: Affective Computing and Intelligent Interaction (ACII).

Ekman, P. and W.V. Friesen., 1978. *Facial action coding system: A technique for the measurement of facial movement.* San Francisco CA: Consulting Psychologists Press.

Ekman, P., W.V. Friesen, and P. Ellsworth., 1972. *Emotion in the human face.* Newyork: Pergamon Press.

El Kaliouby, R., P. Robinson, and S. Keates., 2003. *Temporal context and the recognition of emotion from facial expression.* Crete: HCI International.

El Kaliouby, R. and P. Robinson., 2004. *Real-time inference of complex mental states from facial expressions and head gestures.* Washington DC: IEEE Computer Visions and Pattern Recognition (CVPR).

El Kaliouby, R. and P. Robinson., 2005. *Generalization of a vision based computational model of mind-reading.* Beijing, China, Affective Computing and Intelligent Interaction (ACII), pp. 582-589.

Mahmoud, M., T. Baltrušaitis, and P. Robinson., 2012. *Crowdsouring in emotion studies across time and culture.* Nara, Japan: ACM Workshop on Crowdsourcing for Multimedia.

Marchi, E. et al., 2018. The ASC-Inclusion perceptual serious gaming platform for autistic children. *IEEE Transactions on Computational Intelligence in Games.*

Picard, R. W., 1997. *Affective Computing.* Boston: MIT Press.

Riek, L. D., P. Paul, and P. Robinson., 2010. When my robot smiles at me: Enabling human-robot rapport via real-time head gesture mimicry. *Journal on Multi-modal User Interfaces,* 3(1-2), pp. 99-108.

Riek, L. D., T. Rabinowitch, B. Chakrabarti, and P. Robinson., 2009. *Empathizing with robots: Fellow feeling along the anthropomorphic spectrum.* Amsterdam: Affective Computing and Intelligent Interaction (ACII).

Riek, L. D. and P. Robinson., *Synthesising head and facial movement disorders on android robots.* Edinburgh, Scotland: ACM Symposium on Facial Analysis and Animation.

Robinson, P. and T. Baltrušaitis., 2015. *Empirical analysis of dimensional affect.* Xi'an, China, Affective Computing and Intelligent Interaction (ACII).

Robinson, P. et al., 2011. *The emotional computer.* San Francisco CA: ACM Conference on Pervasive Computing.

Russell, J. A., 1980. A circumplex model of affect. *Personality and Social Psychology,* 39(6), pp. 1161-1178.

Schuller, B. et al., 2015. *Recent developments and results of ASC-Inclusion.* Atlanta GA: ACM Intelligent User Interfaces.

Sobol Shikler, T. and P. Robinson., 2010. *IEEE Transactions on Pattern Analysis and Machine Intelligence,* 32(7), pp. 1284-1297.

Wright, J. et al., 2017. *Intelligent scheduling for in-car notifications.* Modena, Italy: IEEE Forum on Research and Technologies for Society and Industry.

Wyatt, J. and P. Robinson., 2018. *God, artificial intelligence and humans: Could robots replace humans made in the image of God?.* Cambridge: Faraday Institute.

Spirituality, the brain, and religious experience

13.

Psychedelics, implants, spiritual enhancement, and a computational ethical proposal for harnessing spiritually augmenting BCIs

Philip Butler

Loyola Marymount University

Abstract

In this chapter, Butler explores potential directions for spiritually augmenting brain-computer interfaces (SABCIs), and the importance of employing computational ethics for emergent technologies. This chapter looks into the ways that neurological understandings of three psychedelic substances (ayahuasca, psilocybin, and LSD) can form the foundation of SABCI experiences. Here, SABCIs function as speculative transhumanist machinery to engineer spiritual augmentation through the regulation of categorically spiritually experiences. This chapter begins with a neurological exploration of the effects of these psychedelic substances have on the brain. This is done in order to unearth some level of similarity among the chosen substances. It then moves to offer possible SABCI experience profiles, and races questions regarding the ethical distribution and regulation of SABCIs. Finally, this chapter concludes with a proposal for emergent technologies, such as the SABCI, which is grounded in computational ethics.

Keywords: Spirituality, Psychedelics, Neuroscience, Technology, Transhumanism, Ethics, Structural Equation Models

Introduction

This chapter will do two things. First, it will explore potential directions for spiritually augmenting brain computer interfaces (SABCIs), as transhumanist machinery, to engineer spiritual experience by mimicking the biochemical processes of psychedelics. Secondly, it will explore the importance of employing computational ethics for emergent technologies, with SABCIs in mind, as a means to decipher their most equitable use.

Recognizing the ambitious nature of this chapter, I think it is important to outline the five major assumptions which underly it.

First, the exploration of potential SABCI devices is grounded in a non-reductive physicalist approach to spiritual experience. Within this framework spiritual experience is directly tied to human biology, although not confined to the limitations of what is currently known about human biology. The spaces between materiality, human knowledge, and concepts like spirituality should be viewed within the parameters of emergence (Clayton and Davies, 2006).

Secondly, if psychedelics are to be viewed as a potential biological archetype of spiritual experiences, then I think an emphasis on neurophysiology is a necessity when contemplating the replication of neural network activations through SABCIs. Furthermore, psychedelic drugs are chosen over and against spiritual experiences associated with a particular religious or spiritual tradition due to their efficacy in aiding individuals in reaching "categorically" spiritual experiences/spiritual ecstasy. I define a categorically spiritual experience as an experience of high emotional arousal where one's perception of self alters significantly. In doing so, a categorically spiritual experience may encompass experiences associated with a wide range of traditions. This should not be mistaken as an attempt to reduce spiritual experience to mechanical processes. However, it is meant to highlight two points of interest to the discussion: 1) the moment when an individual knows that s/he/they is having a spiritual experience; and 2) the transformational aspect of that spiritual experience. Focusing on the alteration of self allows for broad application as it can be described in terms of non-dualism or unitary experience, nothingness, or even everythingness (Josipovic, 2014). Attention is given to altered perception of self because what one believes about one's self might be key to shifts in individual behavior (Piff et al., 2014), or the outward manifestation of inward transformation. Some might argue this outward manifestation is the overall purpose of engaging in spiritual practices. Categorical spiritual experiences can last for a short or an extended amount of time, and be mediated by any religion, spirituality or psychedelic substance. As such, biological processes facilitating the alteration of perceptions concerning the self can originate from dream states, transcendent experiences, religious practices, meditations, or psychedelic hallucinations.

Thirdly, an assumption of this chapter is that SABCIs will eventually exist. This chapter does not focus on the necessity of SABCIs, the ability of SABCIs to create greater levels of spiritual convenience for their users, the ability of

SABCIs to facilitate connections with the divine, or even allude to the overall utility of SABCIs. This chapter also does not consider whether devices like these ought to exist. Within this chapter SABCIs function as a speculative product of transhumanist modes of technological advancement. They are situated as a potential synthesis of technology, human biology, and spirituality. So, the relevance of exploring SABCIs in relation to human spirituality has more to do with exploring a wider and more customizable notion of spiritual experience. Thus, spiritual augmentation falls in line with transhumanism's framework as a mode of capitalist inevitability rather than an extension of any particular religious or spiritual tradition. The proposed existence of SABCIs can be argued as a viable reality, given recent trends towards greater plurality of religious expression and the widening landscape of spiritual exploration (Pew Research, 2015). While this chapter could explore SABCIs that mimic the neural imaging of meditation experiences, or what may qualify as more traditional modes of spirituality I think it might be more fun to explore spirituality through the neural impact of psychedelic substances, even though the natural form of psychoactive substances are widely available—although through primarily illegal channels (continued in next comment)

Fourthly, this chapter views religion in terms of its etymology, either as that which binds or that which brings together (Derrida and Anidjar, 2001). The idea is that sociality functions as a bridge between individuals. More importantly, this approach widens the definition of what might be considered religiosity, and therefore what might also be considered spirituality—as the manner in which one lives out what one believes. It is an attempt to engage physicalism to demystify spirituality through neurophysiology without reducing the experience of spirituality. Furthermore, it acknowledges the inherent complexity of the religious experience, but does not stop at the door of complexity in wrestling with what might be some of the primary components that contribute to the complexity of spiritual experiences.

Lastly, the exploration of SABCIs, and SABCI ethics presented within this chapter is grounded in the perceived wisdom of predictive modeling. This framework seeks to move towards understanding predictive modeling as a normative conceptualization tool when imagining the distribution of technological advancements. Focusing on the strength of predictive modeling creates space for ethicists to simultaneously imagine the implications of multiple variables for future outcomes in a way that more traditional ethical models cannot. The variables specific to this chapter are the exponential increase in technological innovation and the core components of consideration for the equitable distribution of emergent technologies (i.e., race, class, and gender). And while race, class, and gender would normally suffice for intersectional ethical analysis, factors related to the brain that SABCIs would influence (namely, cytoarchitecture, neurophysiology, and neurochemistry) are also important when considering the regulatory implementation and distribution of SACBIs. The multivariate nature of this ethical framework is meant to allow for the complexity of the ethical decision making process to emerge while presenting the benefits and consequences of the foreseeable paths discussed herein.

This chapter will be broken into several sections. The first section will be an outline of the neurophysiology of three separate psychedelic drugs: Ayahuasca or N,N-dimethyltryptamine (DMT); Lysergic acid diethylamide (LSD); and Psilocybin (Shrooms). Next it will outline a basic synthesis of the key neurological outcomes across those three substances that will allow SABCIs to mimic their neurophysiological effects. Finally, this chapter will construct a computational ethical framework for the production and distribution of emergent SABCIs as a potential template for equitably distributing emergent technologies.

The neurological impact of psychedelics

Psychedelics provide a clear neurophysiological window into the 'categorically' spiritual experiences. Again, this is an inference that spiritual experiences are grounded in materiality, and are experiences had through the vehicle of individual biology. The connection between neurophysiology and spirituality as it pertains to biomimicry has much to do with what the neurosciences currently "know" about the experiences that can be classified as spiritual. Following this knowledge would be an attempt to engineer the biological modulation of those experiences. One could consider transhumanist approaches to human enhancement (via technologies such as BCIs) that allow for individuals to regulate the intensity and duration of a categorically spiritual experience as being laden with the potential to cultivate prosocial behavior. And this bioengineering of spiritual experiences through neurophysiological regulation becomes the basis for spiritual augmentation. Here spiritual augmentation is where human enhancement meets spiritual intentionality. The process one would undergo to enhance one's self would be the preliminary exercise before one might gain the ability to reach the heights of spiritual experience at will, or at a preset time. However, similar to any spiritual experience engineers could not regulate what occurs, on the level of subjectivity, once regional network activations occur. But theoretically, an SABCI might at least be capable of transporting individuals to what some believe to be the spiritual plane.

Individuals who engage in the use of psychedelics overwhelmingly stress the consistent nature of psychedelic affect (Tart, 1972). This is in comparison to religiously mediated spiritual experiences where (whether Jewish, Christian, or

Muslim) individual adherents may, or may not, perceive their own experience as consistently spiritual or emotionally significant. The apparent connection between emotionally significant experiences and spiritual experiences can be understood through the depth of emotional immersion that participants/adherents encounter in correlation to a given spiritual experience. Furthermore, an exploration of emotionally immersive experiences as spiritual experiences, especially when connected to psychedelic substances, may potentially aid in determining what it means, in neurological terms, to be grounded in a particular spiritual experience. This is also important as spiritual events associated with psychedelics do not require 'belief' in psychedelics themselves, readiness/preparation for a psychedelic episode, or the routinization of psychedelic use in order to reach/experience psychedelic efficacy. So, for the purposes of this chapter, a categorically spiritual experience will be defined as an experience of high emotional arousal where one's perception of self alters significantly. The alteration of self can be described in terms of non-dualism or unitary experience, nothingness, or as everythingness (Josipovic, 2014). This experience can last for an extended amount of time, and be mediated by any religion, spirituality or psychedelic substance. As such, processes facilitating the altering of perceptions concerning the self can be grounded in hallucinations, dream states, religious practices, or meditations. However, the intermittent qualities of the categorically spiritual experience precipitate its described elusive subjectivity. In many cases this leads individuals to infer that although spiritual experience is essentially ineffable there is a definite 'knowable' quality, whereby individuals who have a spiritual experience know that they have had one. This is especially valid when an experience fits within the bounds of the categorical spiritual experience described above. The chemical potency of psychedelic substances which allows for the threshold of the categorical spiritual experience to be reached is the primary reason that psychedelics provide a strong case for biomimicry when constructing SABCIs.

Historically, the connection between psychedelics and spirituality has stemmed from the relationship psychedelics have had with indigeneity. For instance, Ayahuasca use has been tied to shamanism found in South America. Shroom usage dates back to prehistoric times in Africa, which is evidenced through cave drawings. LSD is a more recent addition. It was produced in a laboratory, and is efficacious in micro-doses— which might point to its increased potency. Nonetheless, it is well documented that psychedelic prominence within multiple Indigenous spiritual traditions stems from the ability of psychedelics to invoke profound emotionally immersive experiences in conjunction with these various spiritual contexts. Furthermore, rises in indigenous practices in the North Americas might also strengthen the case for psychedelics to from the biological archetype of categorically spiritual experiences. These categorically spiritual experiences have led to a large body of literature covering the neuroscientific underpinnings of the experiences elicited by these substances: neurologically, physiologically, and chemically. So, to explore the potential biological impact of SABCIs mimicking these substances, I will spend the remainder of this section engaging prominent studies of DMT, Psilocybin, and LSD.

DMT

Ayahuasca use has seen a spike in recent years given an increased interest in indigenous practices (Inglehart and Norris, 2003; Tupper, 2009). Users of ayahuasca report categorically spiritual experiences which include: altered somatic awareness, increased emotionality, heightened introspection, and vivid visualizations. Three prominent studies (Riba et al., 2006; Draulio et al., 2012; and Palhano-Fontes et al., 2015) outline the neurological implications of ayahuasca use provide insight into understanding the role regional functionality and connectivity plays in contributing to the spiritual features associated with ingesting ayahuasca.

Riba et al. (2006) conducted a single photon emission computer tomography (SPECT) study to determine regional cerebral blood flow resulting from the use of ayahuasca. Their results highlighted increased activity in areas known to coincide with the brain's emotional network (Schilbach, et al., 2012). SPECT images showed a bilateral increase in blood flow in the anterior cingulate (ACC)/frontomedial cortex (MPFC) of the right hemisphere. Increased blood flow was also present, bilaterally, in the anterior insula, although greater intensity was found in the right hemisphere (Riba, 2006). Increased blood flow appeared in the left amygdala/parahippocampal gyrus (PHG). These results suggest a positive correlation between the emotional effect of ayahuasca and its neural correlates. The relationship between the ACC/frontomedial cortex, somatic awareness, and introspection point to the reports from users who experienced altered states of somatic awareness and increased introspection. The anterior insula and parahippocampal region's role in emotional arousal and self referentiation correspond to the overall heightened sense of emotionality and introspection experienced by ayahuasca users. However, the evidence from the Riba et al. Study does not point to the vivid hallucinations broadly reported by ayahuasca users.

Draulio et al. (2012) performed a functional magnetic resonance imagining (fMRI) study to determine the neurological influence of ayahuasca on the vivid images reported by users. The research team used psychological scales to ground the fMRI images recorded. They presented a series of images to participants. First they presented clear (non-pixelated) images of animals, plants, and humans. Next, the team presented scrambled (pixelated) images, and participants were presented with a visualization task. Observing blood oxygen level dependent (BOLD) activations related to visualizations post- intake of ayahuasca allowed researchers to see which brain regions become more active as participants underwent the study's visualization conditions. Regarding ayahuasca visualization, there were increased activations in the primary (medial occipital lobe or V1) and secondary visual cortexes (V2, V3), frontal cortex (via the ventral prefrontal cortex (VPFC) and orbitofrontal PFC), and temporal regions (superior temporal gyrus and superior and middle frontal gyrus) via their connection to V1. Remarkable to the study's findings were that the activations of closed-eyed visualizations were comparable to the activations in response to viewing clear images. These increased activations were said to be mediated through the connective relationship initiated by the V1, stemming out to secondary visual cortexes leading to the frontal and temporal regions (Draulio et al., 2012). For instance, the medial temporal lobe was activated via the connectivity of the ventral stream of visual processing (Goodale and Milner, 1992). Generalized Ayahuasca use, according to this study, also facilitated increased bilateral activations in the precuneus, lingual gyrus, fusiform gyrus, middle occipital gyrus, parahippocampal gyrus, and posterior cingulate gyrus.

Palhano-Fontes et al. (2015) explored the impact of ayahuasca on the brain's default mode network (DMN) in contrast to the task performance network (TPN). The DMN and TPN have an inverse relationship in terms of brain activity (Fox et al., 2005; Uddin et al., 2009). The DMN is responsible for conceptions of self (often negative), memory retrieval, future planning, and mind wandering (Buckner and Carroll, 2007; Northoff et al., 2007). The TPN is associated with externally-oriented task-based concentration (Fox et al., 2005). Palhano-Fontes et al. Found that ayahuasca use disrupted the resting-state connectivity (rsFC) of the DMN through decreased activation in the PCC/precuneus while maintaining an inverse relationship to the TPN. Resting-state functional connectivity can be described as time elapse correlations between regional activations. So, regions are said to be functionally connected if their activations are temporally correlated. In terms of the rsFC of the PCC, this decreased activity contradicts Draulio et al.'s study which suggests the PCC experiences increased activation. However, this could be due to the specificity of each task being tested. The neurological pathways of episodic self-reference (PCC/precuneus, PHG, MPFC, and temporal regions) is simply different from visualization (V1-4, MPFC, temporal and limbic regions), although there is some admitted crossover especially since thinking of the past and future are often accompanied by mental visual representations. So, it would be safe to suggest that ayahuasca use could down-regulate the PCC during introspective components of the concentration demanded by the experience, while re-engaging the PCC during the hallucinogenic visualizations. It is important to note that the decreased activation of the PCC during ayahuasca use, as references by Palhano-Fontes et al., is in specific reference to the rsFC of the DMN. The decreased activation of the rsfc that is facilitated by the PCC might be a key component of the biological space needed for altering perceptions of self, given the PCC's role in negative self-reference (Cooney et al., 2010). Important to the discussion, support of the longitudinal impact of ayahuasca use on the PCC has been presented by Bouso et al. (2015). They looked at potential cytoarchitectural differences in the brain as a result of long term psychedelic usage, and they found that ayahuasca users had less cortical thickness in PCC. Lower density in the PCC has been correlated with higher perceptions of the self—a positive aspect connected to spirituality (Vago and Silbersweig, 2012).

Psilocybin

Psilocybin produces hallucinations, a decreased sense of self (ego-dissolution), euphoria, changes in consciousness, and distortions of time. Carhart-Harris et al. (2012) published an article that outlined the neural correlates of psilocybin. The researchers analyzed the cerebral blood flow (CBF) and BOLD signals of 15 subjects after ingesting psilocybin. Results demonstrated a global decrease in CBF and BOLD signal. Still, significant decreases in activation were found in primary association/connectivity hubs: thalamus, medial PFC (mPFC), ACC, and PCC. This is important to note, given the role that these structures play in the DMN, i.e., mind wandering and negative self-conceptualizations. Researchers observed that greater decreases in the ACC/mPFC were connected to a greater intensity of the subjective effects of psilocybin (Carhart-Harris et al., 2012). Furthermore, not only were these regions markedly less active, their rsFC was decreased as well. The importance found in the decoupling of the mPFC and PCC suggests that decreased activation in these regions might facilitate a lessened awareness of one's self/body, and a less critical/negative approach to one's perception of self.

The findings of Carhart-Harris et al. do contradict a previous psilocybin study conducted via positron emission tomography (PET). Vollenweider et al. (1997) found that psilocybin increased global glucose metabolism after ingestion. However, Carhart-Harris explained this discrepancy with two possibilities. One explanation suggests that the radiotracer used by Vollenweider et al. to measure glucose metabolism has a long half-life, roughly 110 minutes. This long half-life makes it possible for phasic or short term effects of psilocybin to demonstrate rebounding capabilities detectable by longer-term changes in glucose metabolism. Another explanation draws from Rodrigo Andrade (2011), who explains the neuropharmacological role of psilocybin and its potential interactions with serotonin receptors and pyramidal cells (5-HT2A).

> Stimulation of the 5-HT2A receptor increases excitation in the host cell by reducing outward potassium currents. So, if the 5-HT2A receptor did mediate the observed deactivations, then it may have been via 5-HT2A-induced excitation of fast-spiking interneurons terminating on pyramidal cells, or even 5-HT2A-induced excitation of pyramidal cells projecting onto interneurons (Carhart-Harris et al., 2012: 2141).

Still, the overall findings of Carhart-Harris et al. (2012) point to the importance of the mPFC and PCC in the neuromodulation of the psilocybin. Muthukumaraswamy et al. (2013) conducted a study using magnetoencephalography (MEG). MEG uses small magnets to measure the electric field created by the electric current of neurons (Singh, 2014). In this instance, global decreased activity was observed again. Greater decreases in activation were found in association areas, namely the frontal and posterior association regions respectively. These findings were in some ways a synthesis of Carhart-Harris et al. (2012) and Andrade (2011), as this study also examined the correlation between psilocybin and the 5-HT2A receptor laden pyramidal cells that are dense within the PCC. The findings of Muthukumaraswamy et al. (2013) suggest the decreased neural activity in PCC has a positive relationship with psilocybin's ability to excite the 5-HT2A receptors which then facilitate the decrease in activity and decoupling of the PCC from its association networks. Again, the decreased activity of the PCC, and the decoupling of the PCC from its other associative regions, i.e. the DMN, further align this study with previous literature which suggested that PCC activation has an inverse effect on the intensity of user's experience post psilocybin intake. A more recent study (Lebedev, 2015) further supports the diminished activation of self-referencing areas, particularly the frontal lobe areas, and its connectivity to the medial temporal lobe (MTL) as adding to the loss of self that users experience after ingesting psilocybin. This study points to the importance of the parahippocampal cortex (PHC) and retrosplenial cortex (RSC) in ego-dissolution. Both the PHC and RSC play an integral role in episodic memory and self-reference. Decreases in these regions suggest the rsFC that the PHC and RSC share with the PCC is a key component to the intensity of the ego-dissolving experience associated with psilocybin. These findings are also supported by a study by Kometer et al. (Kometer et al., 2015). This study used electroencephalography (EEG) to further implicate the PHC, specifically, as being part of the ego-dissolution that occurs during psilocybin use—which is the result of the disintegration of the rsFC it shares with the PCC. Psilocybin also shares a similar neurological outcome to DMT, inhibiting PCC activation for less negative self-referential thinking.

LSD

LSD has less historical connection to the categorically spiritual experience since it was created by Swiss scientist Albert Hofmann in 1938. LSD users report hallucinations (vivid visualizations), ego-dissolution, an altered sense of meaning, and changes in consciousness (Hofmann, 1980). It is seen as a much more potent version of psilocybin due to the similarity between their molecular construction, and because LSD can be taken in microdoses. Carhart-Harris et al. (2016) conducted a multimodal imaging study on the neurological impact of LSD. Her research team utilized arterial spin labeling (ASL), fMRI, and MEG as complementary measures to observe resting states after LSD consumption. Prior to this study LSD imaging studies dated back to the 1950s and 1960s. Carhart-Harris et al. saw this study as an opportunity to re-enter the LSD imaging research with more robust measures for a well-rounded investigation. The convergence of these neuroimaging measures showed increased CBF in the visual cortex via fMRI imaging and ASL, decreased visual cortex alpha power via MEG imaging, increased rsfc of V1 via fMRI and ASL, decreased rsFC in the DMN via the PCC as shown by MEG imaging, and decreased rsfc between the PHC and RSC via fMRI imaging. The neurological effects of LSD presented in this study also strongly correlated with the psychological effects reported by LSD users. Increased V1 activations, decreased visual cortex alpha power, and an expanded visual cortex correlated with the intensity of vivid visualizations. Furthermore, decreases in rsFC of the PCC, PHC, and RSC showed a strong connection to rates of ego-dissolution and altered meaning/understanding of self (Carhart-Harris et al., 2016). Although, PHC and RSC disintegrations showed stronger connections to altered states of consciousness.

Neural commonalities of psychedelics

Neuroimaging literature suggests that there are five commonalities significant to the psychedelic experience. Increased activations in V1 that correlated with user hallucinations were observed for DMT, psilocybin, and LSD. However, an expanded visual cortex and increased V1 rsFC were particularly observed in DMT and LSD use. Disintegration of the PHC-RSC connection, along with disintegration of the PHC-RSC connection to the PCC were both found in studies of psilocybin and LSD. A similar activation was found in DMT, where the left amygdala/PHG were seen as exhibiting increased activation and subsequently were implicated in the emotional valence and highly introspective nature of the DMT psychedelic experience. However, the most prominent commonality across the psychedelic experience was the disintegration of the DMN. Decreases in rsFC were found in both the mPFC and PCC. The roles that the mPFC and PCC play in negative perceptions of self and mind wandering under baseline neurological conditions suggests that the altered states of consciousness, connected to the broader psychedelic experience (associated with DMT, psilocybin, and LSD) are heavily influenced by an altered sense of self.

SABCIs recommendations

Creators of SACBIs that mimic the psychedelic experience in order to engineer categorically spiritual experiences must consider several factors. For instance, which psychedelic experience? Or, how long should a psychedelic experience be allowed to last, and at what intensity? Furthermore, questions regarding the impact SABCIs might have on religious affiliation are important, given the capacity of SABCIs to regulate categorically spiritual experiences. If engineers of the future are to build an SABCI based off current neurological knowledge of DMT, psilocybin, and LSD, then I propose they will have one of three primary options: (1) generalized amalgamation of commonalities, (2) profile selection; and/or (3) customized combinations. The remainder of this section will be dedicated to exploring these three options.

A generalized amalgamation would utilize the overlapping formulation alluded to in the previous section that draws from the broad commonalities of neural correlates associated with all three psychedelics. It would then, at varying levels of intensity, increase activation and rsFC in V1 for vivid visualization intensity. Depending on the level of moderated intensity an amalgamative SABCI would also extend the visual cortex into V2, all the way up to V4. An amalgamative approach would also decouple the mPFC and PCC while activating PHC-RSC and decreasing the rsFC of the PCC in order to facilitate ego-dissolution. So, theoretically this amalgamative approach would produce a psychological effect similar to key components of each psychedelic experience.

Profile selection consists of three components that allow for the integrity of one experience to be maintained over against another. First, it would require patients/customers, or end-users, to determine which of the three end users (patients/customers) to determine which of the three end users (patients/customers) to determine which of the three psychedelic profiles of experience they wanted to set their device to (permanently or temporarily). A permanent selection might be the result of a particular type of device, or set of devices that are programmed for one specific psychedelic experience. The second component moves beyond the basic matrix of psychedelic experience that includes the neural correlates of all three substances. The SABCI programmed for profile selection would need to be tuned to the specific intensity of the individual which leads to the third component. The third component of the profile selector based SABCI would be based on individual self-report and neurochemistry. This is where it gets interesting. Depending on whether the end user is a patient or a purchasing customer determines the 'best' outcome of their experience. What are the differentiators of a patient or a customer? For someone to fall under the classification of a customer would suggest that receiving an SABCI is an elective procedure for them. This implies that customers desire to experience hallucinations more frequently, or to reach states of deep introspection and ego-dissolution at will, or at a scheduled time as part of their own techno-regulated spiritual practice. Patients, however, would be faced with the need to mitigate intense psychological episodes, or attenuate maladaptive biochemistry through the therapeutic qualities of an SABCI. Admittedly, more complex scenarios for why someone would want or need and SABCI would arise. So, due to the complex circumstances that might lead one to receive an SABCI, end users would need to undergo preliminary psychiatric testing. These psychiatric tests would help determine previous psychological histories and potentially undiscovered/untreated pathologies via: depression scales (Hamilton, 2004), anxiety scales (Maier et al., 1988); other psychopathology scales (Arbisi and Ben-Porath, 1995), spirituality assessments (Underwood and Teresi, 2002), wellbeing scales (Tennant et al., 2007); altered states of consciousness scales (Studerus, Gamma, and Vollenweider, 2010); and visual analog scales (Miller et al., 2003). The combination of these assessments would help to set a personal baseline for the end user. These self-reported scales would then need to be tested against either actual experience with DMT, psilocybin, or LSD, or through the utilization of the altered states of consciousness scale to determine preferred elements of altered states of consciousness that patients/customers would need, and or want, to be regulated through their SABCI device. Individual neurochemistry, which is the second part of the third component is a bit more difficult to determine. Personalized medical therapies are still in their nascent stages. So, determining the 'correct' dosage of a preset profile would be in response to individual neurochemistry; requiring a feedback loop built into the SABCI itself that is able to subsequently alter the dosage. Neurochemistry regulation is also atypical for the majority of current BCIs, because most of them utilize EEG to communicate with computers.

The customized combination option would allow for patients/customers to select from any variety of the three psychedelic experiences for desired, or needed, purposes. It could allow for all three experiences to occur, at once—as one loud/overwhelming experience. Or, it could allow for selective components from two or more of the different psychedelic profiles to be selected and experienced. For instance, the V1 stimulation of the LSD experience might be fused with the DMN PCC disintegration associated with the DMT experience. It could also include the decoupling of the PHC-RSC connection and left amygdala activation of DMT coupled with the visual cortex activations of the psilocybin psychological experience. It is important to note that along with the various combinations allotted within the blending of these psychedelic profiles that levels of intensity matters as well. The customized combination option would provide different experiences for both the patient and the customer. Patient customization would fall closer to the profile selection with the added choice of having multiple settings based upon therapeutic needs. Customers might have the option of selecting their attributes and changing their psychedelic profiles in real-time. But, the customized combination presents serious problems. For instance, should end users get to decide the kind of experience their SABCI provides, and do they get to control the SABCI themselves?

In 2016, a study was conducted to compare closed-looped BCIs to open-loop BCIs in order to determine end-user comfort while exploring ethical concerns surrounding the device (Klein et al., 2016). These BCI devices were programmed to provide deep brain stimulation (DBS) in order to help participants with major depressive disorder (MDD) and obsessive-compulsive disorder (OCD) alleviate their symptoms. Within the analysis of the data four main ethical themes arose from the individual use of the closed-loop DBS device: (1) control over device function, (2) authentic self; (3) relationship effects, and (4) meaningful consent. These factors were found to be most important to the participants who received a BCI during the course of this study, which suggest some key guidelines for personal BCI therapy regulation. Participants actually did not want control over the device. They felt unqualified to handle it, and preferred to have the physician control the device. Participants were also worried about their ability to maintain the key components of what they felt to be their authentic self with frequent DBS stimulation. The fear participants expressed regarding a loss of self was accompanied by concerns about how their relationships would be affected. No one wanted to be dismissed as being heavily influenced by the device. Nor did anyone want to be the recipient of negative speech which suggested they needed to immediately engage in DBS treatment. Lastly, participants wanted to have the ability to navigate the process of treatment with “effective communication of complex information, help with framing realistic expectations, and attention to cognitive and affective impairments” (Klein et al., 2016, 3). The concerns expressed in conjunction with the DBS BCI present key ethical issues pertinent to the roll-out of SABCIs. Even more so, they begin to scratch the surface of what it means to have a surgically implanted device that is intended to augment end-user psychology. But what are the religious implications associated with SABCIs?

Pew Research has highlighted a seven percent shift in the religious landscape of the United States, from 2007 to 2014 (Pew Research, 2015a). This shift has manifested in an exodus from formalized religion, accumulating in a growing non-affiliated (nones) population. The nones are primarily composed of former Christians. While many religious traditions have seen losses in membership, overall non-Christian faiths still saw an increase in memberships over the same time period (Pew Research, 2015a). It is also important to highlight that the Pew Forum has classified the growth of the nones as a global trend—forecasting significant increases in their global population over the next thirty years (Pew Research, 2015b). In addition, the perception of psychedelic drugs is changing in Western society. People are not only viewing psychedelics more positively, those who actively engage in their use report a greater sense of spirituality and fulfillment than those who religiously affiliate or not (Móró et al., 2011). With this in mind, one might wonder about the consequences of making the effects of psychedelics technologically accessible, if these religious trends do, in fact, continue. Will people choose efficacy over tradition, and the search for divine connection? Will people find SABCIs to be an avenue to facilitate the categorically spiritual experience, negating other less bioavailable options presented through religion? But, what is most at stake regarding SABCIs and religion concerns the neurological manipulation associated with SABCIs which may reduce the spiritual experience, and lead to a less noteworthy perception of spiritual occurrences—as merely the excitation and inhibition of certain neural regions and pathways. Still, some might ask whether an SABCI is even necessary? Can't someone just get access to psychedelic substances whenever they please? Current psychedelic trips last for hours on end. Theoretically, an SABCI would be able to modulate these experiences according to the lifestyle and needs of the recipient. Furtermore, SABCIs would be able to learn from end user brain behavior and over time predict the need for a categorical spiritual experience—at a relevant intensity, and, for an appropriate amount of time.

Still, questions regarding SABCI production, distribution, and maintenance still remain. For instance, who has the right to harvest the data from SABCIs? How will biological data be maintained? Who has the ultimate governing ability, and decision-making authority concerning societal dilemmas regarding SABCIs? And, what are the longitudinal effects of SABCI implanted technology? Addressing all of these issues would surely require more space than what is allotted within this chapter. However, it might be of greater use to begin reflecting upon the type of ethical framework that may be capable of grappling with the dilemmas that emergent technology, like an SABCI, might present.

SABCI ethics and s-risks

SABCIs, and BCIs for that matter, fit within the realm of emergent technology. As such, the consideration of their ethical use and implementation requires an ethical approach capable of keeping pace with emergent technology. But, what might constitute an ethical framework that mirrors the rate of technological advancement while operationalizing multimodal logic for ethical purposes? And, does this framework work to maintain current social hierarchies, or will it be one that works to create a more equitable future, given the rapid advancement of technological rollouts? In this section, I propose that predictive ethics, or computational ethics (Ruvinsky, 2007) be the foundation of ethical methodologies applied to emergent technologies. Drawing from Galileo, computational ethics is a framework that utilizes "the language in which God has written the universe" (Galilei, 1953). Grounded in computational analysis, predictive ethics is thoroughly equipped to evaluate potential suffering outcomes as a survival risk (s-risk) (Ginzburg, Ferson, and AKÇAKAYA, 1990). So, the ability of predictive ethics to account for complex scenarios allows for it to engage in both historical assessment and future prediction.

I chose to highlight s-risks as opposed to risk alone, because risk creates a much more open/vague space where one must determine what is at risk, and for who? Risk, as a factor, operates under the assumption that there is some foundational component in the current reality that allows risk to be evenly weighed/configured across all domains, thereby justifying the furtherance of one entity over another—even if the said entity already has significant advantages over others. This can be observed when ethical scholars utilize the phrase 'all things being equal'. The phrase all things being equal highlights the importance of whatever ethical dilemma presents itself, while theoretically maintaining the basic tenets of the current reality. All things being equal very rarely peers into the possibility that the value given to the current reality is not a value worth maintaining. This is important to consider as I begin to explore the need for dynamic logic and complexity, in terms of utilizing a predictive ethic that does not maintain the current state of reality through the ethical dilemma. Because the predictive ethical model actively seeks to reconfigure the current reality through its methodology. This is something that popular ethical frameworks such as Kantian ethics or utilitarianism are incapable of doing. Both engage in the theorizing of 'all things being equal' and operate from the basic premise that is protects those who exist in full citizenship. Besides the inherent racism that undergirds Kant's philosophy and value theory signal an immutability which does not allow it properly operate when asked to reconfigure modes of existence outside what may be deemed normative (Anderson, 1995).

I want to highlight that risk alone also connotes larger than human non-human entities, eg., businesses or governments, who face risk, or could pose a risk to large populations (Miller, 1992; Clarke and Varma, 1999). While suffering admittedly leads to similar questions (what is suffering, and for who), s-risks can be more readily evaluated in terms of the complex variables which comprise any given ethical moment: race, contexts (cultures, economic statuses, environments, etc.), educational backgrounds, power dynamics, histories/temporalities, access, etc. It is important to note that s-risks, since their inception, have been recognized as the types of risks that are existential in nature. S-risks precipitate extreme suffering, inclusive of and leading to human extinction (Bostrom, 2002). Now, not all technological advances pose the risk of total human extinction. But, what about the extinction level, or existential level, risk that technology's exponential growth (Kurzweil, 2000) poses to significant portions of the global population unable to afford emergent technology? This question becomes more pressing as emergent technology takes the shape of transhumanist augmentations. Admittedly, the quality of life (QoL) that a blu-ray player provided at its introduction is strikingly different from the QoL that SABCIs or genetic engineering might produce. Even more so, what about the rapid erosion/deterioration of QoL that historically marginalized populations would face in tandem with this exponential growth of technology? As an alternative, one could suggest a predominantly agrarian lifestyle for those left behind in the wake of technological progress. But there remain real limitations to the possibility of an agrarian lifestyle when one does not own land, let alone other necessary prerequisites, that would make living off the technological grid possible. So, maybe extinction is slow, or not at all. But in terms of technological advancement and the equitable consideration of all agents, s-risk extinction might benefit from a wider definition. A wider definition of s-risk extinction would allow for it to be seen through the lens of death, and exile. Death is the obvious form of physical extinction where pockets of a species, or its whole, cease to exist. However, exile as a means of extinction suggests that a faction of a species is no longer accounted for, or given access to the resources necessary (and even if these populations are granted access, access is not given in a timely fashion) for physical and psychological survival. Exile as extinction can be imagined as being allowed to walk the same streets as thriving populations, but simultaneously experiencing suffering that results from comparatively lower QoL factors (leading to many undesirable outcomes), active invisiblization, and devaluation. To quickly circle back to the topic of the SABCI, a predictive ethical model would not only require access for individuals across the primary factors of existence alluded to above (bodies, contexts, educational backgrounds, power dynamics, histories/temporalities, etc.), but would require adequate access to the most premium of SABCI services—regardless of whether end users are patients or clients. But if s-risks are integral to the predictive ethical model, what makes the predictive model formidable for the emergent technological landscape?

The primary components of the predictive ethical model are agility and complexity. Agile systems have the ability to account for the complexity of domain-specific ethical conundrums while allowing for fluidity of the weights applied to variables of particular situations. In *The Ethics of Technology: A Geometric Analysis of Five Moral Principles,* Martin Peterson applies geometric evaluations to create multidimensional spaces for determining the right ethical paradigm to apply towards technologically based decisions. While Peterson's framework provides an excellent foundation for how to select a particular paradigm over another in a given situation, I think that questions remain about the long-term benefits and consequences of a particular emergent technology. Applying a multivariate analysis prior to the determination of which ethical category a particular situation falls under is imperative. A multivariate analysis of variance (ANOVA) allows for the simulation/modeling of ethical theory in action.

Alicia Ruvinsky (2007) in "Computational Ethics" proposed that the combination of computational modeling and ethical theory would be useful for simulations regarding ethical scenarios. Computational modeling would be used as a means to simulate ethical theory, modeling either single-agent ethical dilemmas or the reciprocal influence between individual agential actions and larger societal movements—in multi-agent models. Ethical simulations would occur through quantification of ethical theory, i.e., a mathematical equation, created based on multimodal logic to represent a theory's problem, praxis, and implications. In her outline of multi-agent systems she stressed the importance of creating complete agents within a simulated world. In a computational multi-agent ethical model each agent is given a profile pre-weighted with various attributes (represented by the factors of existence). These agents are then loaded with personal ethical models/ideologies that predisposes each agent's decision making towards particular outcomes once the simulation's engine is running. Essentially, Ruvinsky's computational ethics creates a world that mimics the current world, then simulates ethical actions based on interactions and dilemmas presented to individual, or multiple, ethical agents based upon preloaded personal profiles and ethical proclivities. Computational modeling, as presented by Ruvinsky looks to be a remarkable way to assess the benefits, consequences, and even the likelihood of ethical decision making given a particular scenario, and specific kinds of agents. She also acknowledges the downfalls of computational ethics. One limitation of computational ethics she recognized was an inability to account for the fluidity of human nature, or what might preclude changes in moral ideologies over time. Machine learning and artificial neural networks have evolved to include the ability to add dynamic weights to agents (Wang and Xiao, 2010; Wen et al., 2015), but the full complexity of human nature, in terms of human consciousness still evades current technological advances (Kurzweil, 2010). Ruvinsky also stressed the importance of the ethical theory equation being correct. If the equations are incorrect then the theories are inadequately modeled within agents, and their subsequent simulations will be inaccurate. I would also like to add the importance of the data (including the list of ethical agents) being diverse enough to represent the ethical agents in the real world. While Ruvinsky's approach to computational ethics does an excellent job accounting for complexity through computational modeling I think her applications could be broadened to forecast the correlation between s-risks and emergent technologies.

A widened conception of s-risks which accounts for the impact of global events, inclusive of technological breakthroughs, requires the need to conceive of the potential risks that emergent technologies pose. The premise is that once QoL disparities become too wide s-risks then reach the level of extinction through exile as described above. Building upon Ruvinsky's computational ethics model, I propose applying computational modeling to determine the ethical dilemmas associated with emergent technology. However, the difference would not be to simulate the ethical decisions of individual and multi-agents interacting. The computational ethics of emergent technology would create a baseline of multi-agent profiles, compare the baseline against computed QoL markers while taking into account the expected QoL increase that emergent technology is meant to provide agents within the simulated environment. For instance, a transhumanist augmentation such as the SABCI would hypothetically increase the end user's ability to obtain deep introspection, tranquility and reconstruction of personal notions of the self, resulting in an increased psychological QoL. However, these devices would only be initially available to those who are either above a particular socioeconomic threshold, or require its therapeutic capacities—and fit within the criterion of medical testing protocols. Therefore the access, or lack thereof, to an SABCI device presents a serious dilemma for the determining QoL of those who inhabit a world post SABCI release. These dilemmas are very much tied to the factors of existence that are embedded within the social structures of modern societies. The race, sex, gender, class, education, etc., of agents are all impacted by the ethical dilemmas associated with emergent technology. Due to the complex and intersecting nature of societal structures, factors of existence, individual agents, and larger groups it is important to employ computational models that are both multivariate and intended for structural analysis. So, in order for computational ethicists to adequately compute the variance between ethical agents and the variety of factors and structures they engage, computational ethicists should employ structural equation modeling (SEM) to assess the potential s-risks of emergent technology.

SEM computational ethics for emergent technology

The use of SEMs in computational ethics ought to follow basic preliminary SEM analysis protocols outlined by Schumacker and Lomax (2004): Model Specification; Model Identification; Model Estimation; Model Testing; and Model Modification. These SEM basics will be applied to measurements of stability and change, exponential linear regression models, time series models and survival analysis models (Newsom, 2015). A multimodal approach to s-risks through SEM modeling will allow for a stronger s-risk assessment. Each model has the ability to assess the longitudinal impact of multiple variables. However, each does something a bit different in terms of allowing for a full breakdown of the potential social and structural impact of emergent technology. Each contributes to a larger picture that will inform the ethical decision makers (of the governing bodies, businesses, or hospitals) who must decide how these technologies will reach the public—and at what rate. For instance, stability and change modeling will help determine the stability or instability of QoL. Exponential linear regression models will allow for future forecasting of the potential disparities of emergent technologies, taking particular consideration for the exponential growth trends that technology has and will continue to work within. Time series models will be able to parse out the relationship that QoL has with each group in a societal structural model across certain time periods. This helps create more in depth discussion in comparison to the results of the stability and change model or exponential linear regression models. Survival analysis will model the distribution of emergent technology, treating said distribution as an s-risk level event. It will assess whether the insertion of a particular emergent technology, at a given time, positively or adversely impacts multiple populations in a manner that leads to surplus, equilibrium, or extinction (death or exile). I will leave the actual Greek letter statistical function notations to those with greater knowledge of computational analysis.

QoL is a latent variable (an unobservable variable concretized through the combined measurement of observable variables) that needs to be defined in order to begin computational ethics modeling. Now, QoL will be different for each individual agent, and groupings of agents—these represent individual and group subjectivities. An agent loaded with one race, class, educational background, gender, etc., will have a different perspective of what determines QoL that is much different from an agent who has another set of weights altogether. Variance in weight distribution will also occur once weights are applied to represent larger groupings that classify agents by race, class, or gender based upon historical ideological sidings of groups from these categories; giving attention to variance within groups. Here, in a multi-agent simulation, computational ethicists will imagine the perspective of various agents. This is why I also want to strongly emphasize the need to properly weigh agents and groups. These weights go beyond proper representations of real-world agents and groupings in order to look at broader understandings of QoL—within an ethical model that creates a baseline model for QoL. More importantly, proper weights in a computational model that assesses s-risks of certain populations ought to account for historical disparities associated with factors of existence, in order to stress the need for these agent groupings to have access to emergent technologies. Again, the premise is that the lack of access to emergent technologies has the potential to render historically marginalized groupings extinct. Therefore, a QoL baseline must be juxtaposed with historical disparities/marginalization in mind, in order to ethically produce and distribute emergent technology in a manner that does not exacerbate QoL disparities—or lead to s-risk level extinction. But what happens when disparity levels are accounted for, and maintained, but do not lead to s-risk level extinction? At that point ethical models need to be run to determine ways to lessen, and possibly reverse said disparities, through the production of emergent technology, i.e., through jobs, government incentives, premium access based upon positive historical correlations to marginalized statuses. The measures taken to ease and reverse disparity levels would work to computationally build equity into the future. Given this computation approach to ethical theory, it may be easy to see how the number of regression models capable of simulating ethical outcomes might seem innumerable. But, it is important to note that computational ethics, as described in this chapter, is meant to undo social disparity. Through predictive modeling's capacity to determine the utility of emergent technologies, such as the SABCI, computation ethics can account for emergent technologies and the s-risks associated with their release.

References

Alexander, G.M. et al., 2009. Remote control of neuronal activity in transgenic mice expressing evolved G protein-coupled receptors. *Neuron*, 63(1), pp. 27-39.

Anderson, V., 1995. *Beyond ontological blackness: An essay on African American religious and cultural criticism.* Bloomsbury Publishing.

Andrade, R., 2011. Serotonergic regulation of neuronal excitability in the prefrontal cortex. *Neuropharmacology*, 61(3), pp. 382-386.

Arbisi, P.A. and Y.S. Ben-Porath., 1995. An MMPI-2 infrequent response scale for use with psychopathological populations: the infrequency-psychopathology scale, F (p). *Psychological Assessment*, 7(4), p. 424.

Bostrom, N., 2002. Existential risks: Analyzing human extinction scenarios and related hazards. *Journal of Evolution and Technology*, 9.

Bouso, J.C., et. al., 2015. Long-term use of psychedelic drugs is associated with differences in brain structure and personality in humans. *European Neuropsychopharmacology,* 25(4), pp. 483-492.

Buckner, R.L. and D.C. Carroll., 2007. Self-projection and the brain. *Trends in cognitive sciences,* 11(2), pp. 49-57.

Carhart-Harris, R.L et al., 2012. Neural correlates of the psychedelic state as determined by fMRI studies with psilocybin. *Proceedings of the National Academy of Sciences,* 109(6), pp. 2138-2143.

Carhart-Harris, R.L. et. al., 2016. Neural correlates of the LSD experience revealed by multimodal neuroimaging. *Proceedings of the National Academy of Sciences,* 113(17), pp. 4853-4858.

Clarke, C.J. and S. Varma., 1999. Strategic risk management: the new competitive edge. *Long Range Planning,* 32(4), pp. 414-424.

Clayton, P. and P. Davies., 2006. *The re-emergence of emergence.* Oxford University Press, Oxford.

Cleff T, Rennings K.(1999). Determinants of environmental product and process innovation. *European Environment,* 9(5), pp.191-201.

Cooney, R.E. et. al., 2010. Neural correlates of rumination in depression. *Cognitive, Affective, and Behavioral Neuroscience,* 10(4), pp. 470-478.

Derrida, J. and G. Anidjar., 2001. *Acts of religion.* Routledge.

Fox, M.D. et. al., 2005. The human brain is intrinsically organized into dynamic, anticorrelated functional networks. *Proceedings of the National Academy of Sciences of the United States of America,* 102(27), pp. 9673-9678.

Galilei, G., 1953. *Dialogue concerning the two chief world systems, Ptolemaic and Copernican.* Westminster, England: Modern Library Science.

Ginzburg, L.R., S. Ferson, and H. Akçakaya., 1990. Reconstructibility of density dependence and the conservative assessment of extinction risks. *Conservation biology,* 4(1), pp. 63-70.

Goodale, M.A. and A.D. Milner., 1992. Separate visual pathways for perception and action. *Trends in Neurosciences,* 15(1), pp. 20-25.

Hamilton, M., 2004. Hamilton Depression Scale. *Group, 1,* p.4.

Hofmann, A., 1980. *LSD: My Problem* Child. New York: Mcgraw-Hill and Co.

Inglehart, R. and P. Norris., 2003. *Rising tide: Gender equality and cultural change around the world.* Cambridge University Press.

Josipovic, Z., 2014. Neural correlates of nondual awareness in meditation. *Annals of the New York Academy of Sciences,* 1307(1), pp. 9-18.

Klein, E. et. al., 2016. Brain-computer interface-based control of closed-loop brain stimulation: attitudes and ethical considerations. *Brain-Computer Interfaces,* 3(3), pp.140-148.

Kometer, M. et. al., 2015. Psilocybin-induced spiritual experiences and insightfulness are associated with synchronization of neuronal oscillations. *Psychopharmacology,* 232(19), pp. 3663-3676.

Kurzweil, R., 2000. *The age of spiritual machines: When computers exceed human intelligence.* Westminster, England: Penguin.

Kurzweil, R., 2010. *The singularity is near.* London, England: Gerald Duckworth and Co.

Lebedev, A.V. et. al., 2015. Finding the self by losing the self: Neural correlates of ego-dissolution under psilocybin. *Human brain mapping*, 36(8), pp. 3137-3153.

Maier, W. et al., 1988. The Hamilton Anxiety Scale: Reliability, validity and sensitivity to change in anxiety and depressive disorders. *Journal of Affective Disorders*, 14(1), pp. 61-68.

Miller, K.D., 1992. A framework for integrated risk management in international business. *Journal of International Business Studies*, 23(2), pp. 311-331.

Miller, S.D. et. al., 2003. The outcome rating scale: A preliminary study of the reliability, validity, and feasibility of a brief visual analog measure. *Journal of Brief Therapy*, 2(2), pp. 91-100.

Miranda, R.A. et al., 2015. DARPA-funded efforts in the development of novel brain-computer interface technologies. *Journal of neuroscience methods*, 244, pp. 52-67.

Móró, L., et. al., 2011. Voice of the psychonauts: Coping, life purpose, and spirituality in psychedelic drug users. *Journal of Psychoactive Drugs*, 43(3), pp. 188-198.

Müller-Putz, G.R., R. Scherer, G. Pfurtscheller, and R. Rupp., 2005. EEG-based neuroprosthesis control: a step towards clinical practice. *Neuroscience letters*, 382(1-2), pp. 169-174.

Muthukumaraswamy, S.D. et. al., 2013. Broadband cortical desynchronization underlies the human psychedelic state. *Journal of Neuroscience*, *33*(38), pp. 15171-15183.

Newsom, J.T., 2015. *Longitudinal structural equation modeling: A comprehensive introduction*. New York: Routledge.

Northoff, G., et. al., 2006. Self-referential processing in our brain—a meta-analysis of imaging studies on the self. *Neuroimage*, 31(1), pp. 440-457.

Pandarinath, C., et. al., 2017. High performance communication by people with paralysis using an intracortical brain-computer interface. *E-life*, 6.

Peterson, M., 2017. *The Ethics of Technology: A Geometric Analysis of Five Moral Principles*. Oxford University Press.

Pew Research Center., 2015a. America's changing religious landscape. Pew Research Center. Available at: <https://pewrsr.ch/2xuWo0e> [Accessed 5 July 2019]

Pew Research Center., 2015b. The future of world religions: Population growth projections, 2010–2050. Pew Research Center. Available at: https://pewrsr.ch/2xAvtQt [Accessed 5 July 2019]

Piff, P.K., P. Dietze, M. Feinberg, D.M. Stancato, and D. Keltner., 2015. Awe, the small self, and prosocial behavior. *Journal of personality and social psychology*, *108*(6), p.883.

Pineda, J.A. et. al., 2008. Method and system for a real time adaptive system for effecting changes in cognitive-emotive profiles. U.S. Patent 7,460,903.

Riba, J., S. Romero, E. Grasa, E. Mena, I. Carrió, and M.J. Barbanoj., 2006. Increased frontal and paralimbic activation following ayahuasca, the pan-Amazonian inebriant. *Psychopharmacology*, *186*(1), pp.93-98.

Ruvinsky, A.I., 2007. Computational ethics. In: M. Quigley. Encyclopedia of information ethics and security. Hershey, PA: IGI Global, pp. 76-82.

Schumacker, R.E. and R. G. Lomax., 2004. A beginner's guide to structural equation modeling. New York: Routledge. Singh, S.P., 2014. Magnetoencephalography: Basic principles. *Annals of Indian Academy of Neurology*, 17(1), p. S107.

Studerus, E., A. Gamma, and F.X. Vollenweider., 2010. Psychometric evaluation of the altered states of consciousness rating scale (OAV). *PloS one*, *5*(8), p.e12412.

Tan, D., and A. Nijholt., 2010. Brain-computer interfaces and human-computer interaction. In: B. Graimann, A. Bernhard, Z. Brendan, and G. Pfurtscheller, eds. *Brain-Computer Interfaces.* Springer London. pp. 3-19.

Tart, C.T., 1972. *Altered states of consciousness.* Oxford, England: Doubleday.

Tennant, R., et. al., 2007. The Warwick-Edinburgh mental well-being scale (WEMWBS): development and UK validation. *Health and Quality of life Outcomes*, 5(1), p. 63.

Tupper, K.W., 2009. Ayahuasca healing beyond the Amazon: The globalization of a traditional indigenous entheogenic practice. *Global Networks*, 9(1), pp. 117-136.

Uddin, L.Q., et. al., 2009. Functional connectivity of default mode network components: correlation, anticorrelation, and causality. *Human brain mapping*, 30(2), pp. 625-637.

Underwood, L.G. and J.A. Teresi., 2002. The daily spiritual experience scale: Development, theoretical description, reliability, exploratory factor analysis, and preliminary construct validity using health-related data. *Annals of Behavioral Medicine, 24*(1), pp. 22-33.

Vago, D.R. and S.A David., 2012. Self-awareness, self-regulation, and self-transcendence (S-ART): A framework for understanding the neurobiological mechanisms of mindfulness. *Frontiers in human neuroscience*, 6, https://doi.org/10.3389/fnhum.2012.00296.

Vollenweider, F.X et. al., 1997. Positron emission tomography and fluorodeoxyglucose studies of metabolic hyperfrontality and psychopathology in the psilocybin model of psychosis. *Neuropsychopharmacology*, 16(5), p. 357.

Wang, L. and F. Xiao., 2010. Finite-time consensus problems for networks of dynamic agents. *IEEE Transactions on Automatic Control*, 55(4), pp. 950-955.

Wen, G.X. et. al., 2015. Neural-network-based adaptive leader-following consensus control for second-order non-linear multi-agent systems. *IET Control Theory and Applications*, 9(13), pp. 1927-1934.

Wolpaw, J.R. et. al., 1991. An EEG-based brain-computer interface for cursor control. *Electroencephalography and clinical neurophysiology*, 78(3), pp. 252-259.

Whitehead, A.N., 1978. *Process and reality.* 1929. Corrected Edition. D. R. Griffin and D. Sherburne, eds. New York: Free Press.

14.
Rights and guidelines for protecting cognitive liberty in the age of neuro-engineering

Alan Weissenbacher

Center for Theology and the Natural Sciences at the Graduate Theological Union

Abstract

As one discovers the neural correlates of religious and political beliefs, emotions, and behaviors, it becomes possible to manipulate them by modifying the brain physically, electrically, and chemically. I consider some problematic issues surrounding this project, particularly a lack of definitional clarity and that of the brain's multifunctionality, and I discuss several rights that can protect an individual's psychology and prevent future harms. I analyze how laws protecting cognitive liberty are lacking and in need of revision. I also propose an internal guideline for the field of cognitive engineering that I term 'acceptability across ideologies', so as to protect against coercive and harmful technologies. If researchers are uncomfortable with the idea of their ideological opponents using their discoveries or advances, then they implicitly know that their technology has the potential of abusing others and, thus, should be modified.

Keywords: Alteration of Religious Beliefs, Cognitive Liberty, Cognitive Engineering, Moral Enhancement, Neuroethics, Neurolaw, Neuromodulation

Introduction

In a recent paper published in Zygon (Weissenbacher, 2016), I detailed several possibilities for the neural correlates of dramatic religious conversions in short time frames, such as the Biblical account of Paul of Tarsus when he converted from an enemy of early Christianity to a zealous adherent. I received feedback from a post-doc at Stanford saying, "You have just opened up a research program into influencing and even changing people's religious beliefs electrochemically." His thoughts immediately went to how he could "improve" society by removing various belief systems he thought were problematic, even on unwilling subjects. This need not be limited to religious belief systems. Political beliefs are susceptible to alteration as well (Holbrook et al., 2016).

We are already witnessing the ability to influence religious beliefs electrochemically. A UCLA study revealed that transcranial magnetic stimulation (TMS) can lessen a person's belief in the divine (Holbrook et al., 2016). Various scientists are working on influencing the brain with sound waves, accomplishing the same thing as TMS but with greater accuracy and allowing less proximity to the subject (Singer, 2009). Berkeley neuroscientist Walter Freeman advanced some preliminary ideas on using brain plasticity to alter various religious and political beliefs (Freeman, 1995). Additionally, although at the genetic level instead of the neural, philosopher Mark Walker (2009) has proposed a multi-disciplinary Genetic Virtue Project (GVP) to discover the means of altering a person genetically to increase their capacity for moral action, and moral reasoning and religion are often intertwined.

In the past, the soul was deemed the province of religion and outside the purview of science. However, as we discover the neural correlates of religious experience and morality, such discoveries allow scientific exploration of not only the biological mechanisms but also their manipulation. Knowing the neurological correlates of various thoughts, beliefs, behaviors, and feelings opens up the possibility of modifying these through the alteration of the physical or chemical substrate of the brain on willing and unwilling subjects. This possibility raises the important ethical question: What are the ethics regarding the acceptability of certain types of enhancements or interventions?

In this chapter, I advance areas of reflection and discuss several individual rights to safeguard a person's psychology that ought to be applied before the current nascent technology regarding neuro-engineering religion and morality becomes fully emergent. I will specifically look at altering someone's psychology in the religious and moral domains. I will not address cognitive enhancement, brain-computer interfaces, or issues relating to brain scans, mind reading, and privacy, viewing these as separate, albeit related domains, and ones on which there are already ethical debates in the literature. I begin by examining several areas within the research where refinement is needed for more effective experimentation and to avoid improper implementation that might cause physical or psychological harm.

Definitions

First, when it comes to neuromodulation, what exactly is someone modulating? Definitions regarding targeted intervention are typically poor. A prime example is an experiment conducted by researchers Andreas Bartles and Semir Zeki (2004) where they claimed to have located the neural correlate of romantic love. They studied 17 subjects who reported to be in love, showing them pictures of their romantic partners, and then compared their brain activity with that from viewing pictures of non-romantic partners. Viewing their romantic partners resulted in activity changes in specific brain areas compared to the control.

However, what exactly is romantic love? Is it the same for everyone? Romantic love can be associated with various mixtures of commitment, sexual attraction, trust, familiarity, and friendship. Which one of these were they testing? How does one define love with its various components and expressions so as to claim that a particular region of the brain is associated with it?

Definitions should be as specific as possible when attempting to research the locations of mental phenomena. There is currently no agreed-upon mental taxonomy in the sciences. Each researcher defines their terms in their own way, and most appear to uncritically assume a shared definitional agreement. And just like with the Bartles and Zeki (2004) study, browse a selection of neurological studies on love and you will find the term used but rarely defined with any specificity.

In another example, if one is attempting to scan for lying, one must define what a lie is—is it a white lie, a lie one has rehearsed or had to spontaneously originate, self-deception, a big lie, a small lie, a slight exaggeration, a 'just kidding' lie for fun, or a lie of omission? Or as alleviating depression is popular in neuro-modification research, is there a risk of defining depression so broadly so as to include the entire populace, running the risk of creating a new epidemic akin to that currently found with opioids? Clinically diagnosed depression is one thing, while grief in response to loss is another. What about anticipatory grief, the shared grief that comes from empathizing with another, or the various stages of grief? There are situations where sadness is appropriate. Grieving is an important and necessary part of loss, coming to terms with it, healing, and developing emotionally into a new stage of life. If grief is unresolved or stunted, it can interfere with the ability to empathize, make new attachments in the future, or adjust to a new life stage. Perhaps people need to learn to effectively mourn and be emotionally supportive of others well rather then attempting quick fixes for negative emotions, which might occur if neural therapies for depression are not tightly regulated as to what types of depression are appropriate for the therapy and those who are able to use the therapy.

Applying this definitional concern to the neurological study of religion, when someone claims to find the neural correlates of religion, what do they mean? Neural correlates are those areas of the brain which are active during an experience, action, or emotion. Locating the neural correlates of religion began with a study by neuroscience pioneer Vilayanur Ramachandran (1998) where he discovered that religious words elicited a high-level physiological response in individuals with temporal-lobe epilepsy, indicating that the temporal lobe might be associated with religion. Not long after, Michael Persinger (1983) reported that he was able to artificially create religious feelings in people through a magnetic field that stimulated a person's temporal lobes. Later studies of Tibetan monks during meditation and Carmelite nuns in prayer revealed that both evidenced similar changes in brain activity (Beauregard and Paquette, 2006; Newberg, 2001). These early studies provided impetus to a research program into locating neural circuits associated with religion

When examining these studies, however, many appear to operate with a singular idea of religious or mystical experience. However, authors have outlined many different forms. For example, Carolyn Franks Davis (1989) provides a taxonomy of various forms of religious experience: an interpretive experience where one gives an everyday occurrence or coincidence a religious interpretation, a numinous experience where one feels insignificant before God's majesty, a regenerative experience where one senses a reception of strength, comfort, or joy, a unitive experience where one feels united to the wider world or cosmos, revelatory experiences or sudden insights that appear to come from elsewhere, or some form of vision of a supra-mundane reality perhaps received through the senses or an internal impression. These definitions are not mutually exclusive, and a person's religious belief system may contain various elements of the above in differing degrees, not to mention the many and varied religious doctrinal commitments a person may have as well.

A large variety of religious experiences are often conflated into a single type. Careful philosophical (and theological reflection when dealing with religion) is needed in order to clearly define the object of study and interpret the results. Greater specificity is needed so one is not attempting to alter something poorly defined. Without specificity in the research of neural correlates, the field of neuro-modification and alteration may proceed in a naïve, unaware, and haphazard fashion, with some possible consequences described later including unintended side effects and accidental alteration of other brain processes.

Functional interconnectivity

Second, I find that some of the literature on neuro-modification appears to operate under the theory of cognitive localization, which attempts to localize complex mental activity to specific, singular, and local areas of the brain. It is true that there are areas of the brain with specific functions, however, many functions of the brain are more systemically distributed and are functionally interconnected. Philosopher of science William Uttal (2001) notes that one must distinguish between a nonhomogeneous brain where different regions interact to influence different mental or behavioral processes, and the hypothesized role of these regions as the unique locations of the mechanisms underlying those processes.

Additionally, many specific brain areas often serve more than one function, yet it is assumed in some studies that once a segment of brain real estate is identified with a particular cognitive process, it is assigned to only that process. For example, there is an area of the brain that has long been considered as responsible for face recognition, yet now it seems to be involved with familiarity of all kinds (Gauthier et al., 2000). Even areas typically thought of as dealing exclusively with sensory or motor tasks are being shown to have considerable overlap and are involved in other functions, even cognitive ones (Stein and Meredith, 1993; Thompson, 1990). Finding a brain region that activates during the performance of a task does not mean that this area operates only to perform this function, and it is possible that in a different context or with a slightly varied task, the same operation could be correlated with an activation in an entirely different brain region. Brains also have many redundant systems for achieving the same tasks. Addressing religion in particular, neuroscience researcher Dimitrios Kapogiannis (2009) describes religion containing interrelated elements of emotion, personal experience, doctrinal beliefs, memory, semantics, cultural influences, and goal-setting. These complex items involve multiple, often different, areas of the brain. The various aspects and types of religious beliefs and experiences are likely whole-brain processes, and research into neural correlates of religion should keep this in mind. Attempting to localize religion to a specific area and attempting to modify such would be like pulling on a single strand of a spider web and being ignorant of the effects this will have on the entire web.

Brain modification has the potential for adverse effects given the brain's multi-functionality. For instance, if you target an area of the insula that deals with drug cravings to reduce or eliminate them or try to modify the area that deals with a sense of union with God, these areas are also correlated with a host of emotional processes. Say a particular neural circuit is responsible for 10 different functions. A researcher targets one for modification, but what will that do to the other 9? For example, even though a modification technique such as deep brain stimulation (DBS) has proven effective in treating various conditions, it has also been shown to cause disruption of one's self-conception of personal identity (Mackenzie, 2011), (Klaming and Haselager, 2013), increased aggression, and impulsivity (Frank et al., 2007), changes in sexual behavior (Houeto et al., 2002), and personality alteration (Pham et al., 2015). When addressing rules surrounding neuro-modification, one should aim to prevent disproportionate harm compared to the therapeutic benefit. At a minimum, brain modification techniques should be required to disclose potential side-effects in the same manner as medical drugs.

Current status of technology

After discussing the challenges and pitfalls of brain modification, is the concern of neuro-chemically modifying religious beliefs on unwilling subjects needlessly alarmist when examining the current status of 'conversion' technology?

Means of modulating neurochemistry already exist. Various types of psychological medications that regulate brain chemistry need not be mentioned as they are fairly common in today's society. Once one knows the neural correlates of a particular psychopathology, it becomes possible to alter it with brain intervention. And while people are generally accepting of brain intervention to address debilitating conditions such as clinical depression or psychosis, one can also intervene in various emotions, beliefs, or behaviors that one would not classify as pathological once their neural correlates are known.

Some neural correlates of religious belief have been discovered (Kapogiannis, 2009), and it is possible that we may discover correlates of religious conversion. While dramatic religious conversions cannot be predicted in advance so as to have a scanner in place at their occurrence, one can at least surmise several possible neurochemical pathways that may be correlated with dramatic conversions. And, as stated earlier, once one knows the neural correlates, intervention becomes possible. Possible pathways for dramatic religious conversions include the hypothalamus, pituitary, adrenal axis which releases globally acting neuro-hormones in situations of stress, modifying the brain in a top-down fashion, or bottom-up processes such as rapid state changes or reorganization after a neurological system is pushed outside of equilibrium (Weissenbacher, 2016). Could it be possible to artificially induce a dramatic religious conversion?

So, we have discovered various neural correlates of religious beliefs, but do we have the technical means to modify these correlates neuro-chemically so as to change them? We already have portable, easy to use devices to administer electric or magnetic stimulation, marketed as a means to boost brain performance or provide relief from migraines depending on the area stimulated. However, switching stimulation to a new area, researchers at UCLA discovered in 2015 that applying magnetic stimulation to the posterior medial frontal cortex reduced someone's belief in God (Holbrook et al., 2016). In their words, religious beliefs are susceptible to targeted neuromodulation, and in the conclusion to their paper published in the journal *Social, Cognitive and Affective Neuroscience*, they advocate for experimental intervention into high-level attitudes and beliefs. High-level beliefs and attitudes include what one factually believes religiously and politically as well as the emotions attached to these beliefs. For example, one may believe in God, but feel positive or negatively about God. Or one may assent to certain religious doctrinal commitments, but view church involvement with negativity. It appears that the Holbrook study advocates a research agenda into the potential to change someone's emotions attached to various religious and political beliefs as well as changing the beliefs themselves through brain intervention – induce someone to love or hate God, or remove belief in God entirely, or perhaps change political affiliations and how he or she will vote in the next election.

While this is an oversimplification, could a person (or 3rd party hacker) turn the switch on someone's portable magnetic brain stimulator from 'migraine relief' to 'become atheist'? This is not far-fetched, however, as a recent 2016 study by Pycroft revealed that consumer-grade transcranial direct current stimulation has few safeguards to prevent third-party manipulation, and brain implants like deep brain stimulation are vulnerable to attack by third parties who want to exert malicious influence over someone's brain activity. Alteration of impulse control, modification of emotions, induction of pain, and modulating someone's reward system can be achieved.

While it seems self-evident that certain companies might desire to manipulate people into various consumer-related behaviors, is there a will to perform these sorts of modifications in the religious sphere, warranting concern? In terms of eliminating religion neurologically, one need only look at the anti-religious polemics found in New Atheism to find religion blamed for a significant portion of the world's ills and opinions on how much better off the world would be without religion. For example, historian Lynn White (1967) is often read as blaming Christianity for the present-day ecological crises and alludes to the idea that saving the earth may only be possible if Christianity is eliminated from the public sphere and replaced with religions that sacralize nature. This provides a strong motive for people to change another's religious beliefs.

Currently in the mental health field one can intervene in a person's psychology against the person's will provided one believes this person is a danger to self or others. And when one deems a particular religion a danger to society, the door opens to the possibility of intervening against someone's will, and as I discuss in the next section, current laws protecting mental integrity allow for such intervention. What makes this particularly problematic is which religion is singled out as a danger will then depend on who is in power. Some see elements of Islam as a danger. Some progressives would classify various conservative Christians as a danger. The New Atheists view any religion a danger. And differing state powers classify particular religions as dangerous, such as China's enmity to the Dali Lama or some of the anti-Islamic rhetoric found in various American or European political circles. It is not hard to envision a particular state forcing religious conversions to obtain a population favorable to the existing power structures. Even presidential candidate Hillary Clinton said at the 2015 6th Annual Woman in the World Summit that certain religious beliefs must be changed. In context she was referring specifically to beliefs opposing abortion, and she left the question open how to accomplish such forced change. Or to flip this concern around, if one is engineering religious belief instead of eliminating it, one need only look at many religions' desire for converts to see how this technology could be appropriated to serve those interests.

Rights and laws

Bioethicist Marcella Ienca and professor of law Roberto Adorno (2017) have proposed some initial regulations for the emerging field of neurotechnology as a whole, which includes issues related to scanning the brain, reading thoughts, performance enhancement, treating neurological disorders, neuro-marketing, and brain-computer interfaces. I describe some of their proposals and modify their suggestions with some additions to address specific issues relating to the neural-alteration of religious beliefs, morality, and related emotions on both willing and unwilling subjects.

Cognitive liberty

The first important law related to the emerging field of neuromodulation is that of the right to cognitive liberty. Neuroethicist Jan Bublitz (2013) explains how this right consists of two related principles: the first that persons have the right to use or refrain from using emerging neuro-technologies, and second, that people deserve protection from coercive and unconsented use of such technologies. A person has the right to mental self-determination, defined as the right to change his or her own mind and choose the means by which this change occurs. A key concept in this right is that of consent. However, consent itself is tricky to define. An inmate offered a lighter sentence may consent to moral modification, but the situation of incarceration itself is coercive.

Informed consent may not take into account whatever situational, interpersonal, or societal pressures one may be experiencing which influence one's decisions toward a particular choice. For example, a child may "willingly" consent to neural engineering to ensure life-long religious commitment or atheism due to home pressures.

Current laws on informed consent stem from the *Nuremberg Code, Rule 1*, which states, that a person should be able to exercise the power of choice free from any element of "force, fraud, deceit, duress, overreaching, or other form of constraint or coercion," as well as "have sufficient knowledge and comprehension of the subject matter involved," so as to make an enlightened decision (Mitscherlich and Mielke, 1949). A person must be able to consent to or decide against participating and have sufficient information so as to make an authentic choice. For example, a doctor must provide information regarding the benefits and drawbacks of a procedure. If the doctor only provides the benefits, then this is coercing someone into making an affirmative choice.

The basic justifying principle and core concept is the protection of autonomy. According to ethicists Beauchamp and Childress (1983), there is a moral duty to "seek valid consent *because* the consenting party is an autonomous person" (68). When a violation of autonomy is institutionally condoned, the right to make such choices is impaired in general. Everyone is placed at risk and everyone stands to suffer. Violating autonomy for a good reason can allow for its violation for a negative one in a different context or situation. If you do not prevent the small violations, large ones may start to occur.

What about principles other than autonomy to protect one from unconsented modification? A principle of utility would justify modification on unwilling subjects when such appears to maximize social welfare, which is often a position taken by moral modification proponents. For instance, Oxford philosopher Julian Savulescu and co-author Ingmar Persson (2008) advance that moral modification is the only means by which to save the planet, and a principle of utility would justify forced modification in this instance. The principle of 'do no harm' (nonmaleficence) could also protect against not seeking consent, however, forced modification can be justified when no apparent harm would result, and especially if the modification appears to be one that would improve a person's life. And who decides what constitutes 'life improvement' for someone else? Considerations other than autonomy have cases that could justify not seeking consent. Persons, however, have rights independent of social utility and risk to subjects that warrant protection.

Further exploration of consent in the area of modification is warranted. To further problematize the issue of consent, intrusions into and modification of one's brain may not involve forcible coercion, as it could be done under the threshold of a person's awareness—while someone is asleep for example. A negative proscription against coercion is not enough. A positive one necessitating enthusiastic and informed consent is warranted.

The right to mental integrity

The second right is the right to mental integrity, which is typically defined as the right to mental health. This right is not a new one and is addressed in the European Union's *Charter of Fundamental Rights*, which states that one has the right to mental health in addition to freedom of thought, religion, and the freedom to change religion (Article 10.1). However, this was drafted before the advent of brain modifying technologies, and this charter is in dire need of revision so to as to take these into account. The 'right to mental health' may not prevent unwarranted intrusion into someone's mental domain and may in fact encourage it in the name of someone's psychological well-being. If I believe that someone's mental health would be vastly improved through a medication he or she refuses to take or a brain alteration, then why not force it in the name of this person's right to mental health? Already one can forcefully commit someone to evaluation at a mental hospital if one believes another is a danger to self or others, or one may give medication without consent if the person is deemed mentally unable to provide such consent. Would not a depressed person be better off having that depression medically removed? And in the midst of such depression, he or she likely is unable to properly consider the benefits of medication or a brain procedure or even be motivated to find relief or take medication on his or her own, and, therefore, one might force these on this person.

Adding to this is the *European Convention on Human Rights* where it states that rights protecting mental integrity as well as religion admit some restrictions, specifically "in the interests of public safety, for the protection of public order, health or morals, or for the protection of the rights and freedoms of others" (Article 9). This statement appears so broad as to allow a whole range of neural modifications on unwilling subjects. For example, who defines what morals need protecting or which should be advanced or curtailed for public safety? Such could vary greatly depending on those in power. This also appears to give license to any state power to curtail certain religions in the 'interest of public good'. The right to freedom of thought may even allow the interpretation that permits one to think, but only with the thoughts given to someone by an outside party. It is a fine line between moral enhancement for the public good and wholesale political re-education.

Already one sees substantial support in the literature for the moral modification of those society deems a danger to self or others. One example is morally modifying violent criminals so that they are safe to return to society. They return to being productive citizens and society saves on prison costs. This example may not be terribly controversial in the literature, yet where does one draw the line and ensure that the line stays there? What about modifying people who get traffic tickets in addition to violent criminals? They endanger society through the possibility of accidents. What about modifying your political opponents? Do not their views endanger society? Does your political opponent hold that your beliefs endanger society?

Some authors also add that modifying certain individuals is a moral obligation even without consent. Returning to the Savulescu and Persson (2008) example, they argue that if safe and effective means of moral enhancement are developed, they should be compulsory for all and may be the only means by which to ensure the long-term viability of humanity. Advancing that forced moral modification is the only means for species survival provides a strong motivation to begin this process as soon as conceivably possible.

I encourage a pause in what I see as an enthusiastic rush to morally enhance individuals so that reflection can occur over the ways this can be abused or co-opted into various political, social, or religious agendas. Various authors in the field appear to be of the mindset that their morality, their worldview, and their personal political or religious concern will be the one people are modified toward. Someone will be the arbiter of what morality to enhance, and it may not always be that with which one is comfortable or may agree with.

Problems will exist even among those who approach moral enhancement with the utmost of good intentions as indicated by philosopher Nicholas Agar (2010) who highlights what counts as enhancements according to one view of morality can be diminishments according to another. Different moral theories highlight different aspects of human nature and moral reasoning as the most important. For example, Kantians emphasize practical rationality, and Utilitarians value minimizing suffering and maximizing pleasure among the most people. Quoting Agar, "Someone who has been subjected to moral enhancement is likely to have reduced sensitivity to moral reasons rejected by his or her enhancer" (75). Agar provides the example of someone considering firebombing a city during a time of war. If such would hasten the end of the war and thus result in fewer deaths, a utilitarian might find firebombing to be the correct action. However, a different value system might reject firebombing, reasoning that the innocents of the city should never be treated as 'mere means' regardless whether more innocents might die in the long term should the city be left alone. In another example, is courage nonviolent resistance to evil or the courage to risk one's life in the military to fight evil? The field of modifying people morally through targeted brain intervention would benefit from a standard of virtue and vice with which to guide the project.

The right to psychological continuity

Since the "right to mental integrity" is not enough to ensure the psychological protection of the individual, the "right to psychological continuity" should be included to address the shortcomings of the right to mental integrity. What this right adds is that it not only protects one from harm, it also protects one from interventions that others deem beneficial or in your best interest (or society's interests). This right preserves the continuity of one's thoughts, emotions, preferences, and related behaviors through time, by ensuring that any changes are not due to external modification of neural functioning. More than protecting mental states themselves, this right protects the neural functioning that produces the mental states. One is free to develop one's own personality and maintain that personal identity without unconsented neural intrusion into the process.

It is true that the rights suggested overlap but taken together they provide well-rounded protection for one's mental domain, not only from harm, but from unconsented intrusion, harmful or otherwise, ensuring genuine informed consent, and curtailing abuses that can result from modifying unwilling individuals if one thinks it is their or society's best interest.

Acceptability across ideologies

The final principle that I propose is not at the level of law, but rather an internal criterion that can serve as an ethical guide to those actively working within the field of neuro-technology and neuro-modification – a principle I term 'acceptability across ideologies'. If one is developing neuro-technology, this person should ask, "Would I be comfortable with this technology in the hands of my ideological opponent, least favorite politician, or someone with a different and opposing religious worldview?" If someone cannot answer yes to that question, he or she is likely developing something that can be used in an abusive manner and should modify the technology or process.

I continue to examine Savulescu and Persson (2008) argument where they discuss the urgent need for moral bioenhancement, citing the need to forcefully modify people to be receptive to and motivated to practice principles of ecojustice. There is not enough time to save the planet to allow for only convincing people through traditional means of persuasion. Peoples' minds must be changed even against their will. I admit that such brain modification might be possible given further advances. The Holbrook study (2016) already indicated that people reflect more pro-immigrant sentiments after the application of targeted TMS, revealing the potential for further and more specific modification of political beliefs. However, using my proposed principle of 'acceptability across ideologies', Savulescu and Persson should ask the question, "Are we comfortable with this technology in the hands of our least favorite politicians?" This technology might not be used to motivate people toward ecojustice. Once one has the technology and lays the groundwork to allow forcible brain change for the purpose of ecojustice, the door is then open to allow for forcible brain change for the purpose of advancing political positions one might find abhorrent—at least abhorrent only until the brain change commences. Then one might find them desirable. Savulescu and Persson might even ask the question about their argument itself. Am I comfortable with a position I disagree with making this argument for forcible brain change? "This particular people group or particular religion represent such a threat to society that there is no time to convince people to oppose them of their own free will, and thus they must be forced to change their opinion through targeted neuromodulation." Savalescu and Persson might then be more hesitant to advocate for the violation of people's cognitive liberty. I use these examples to show that not only should the principle of one's advances being acceptable to opposing ideologies guide innovation, but we should develop laws to protect cognitive liberty as discussed earlier. Violating cognitive liberty for a positive purpose can allow for its violation for a negative one in a different context or situation.

Conclusion

I hold that the ethical issues surrounding the emerging advances in neuro-technology should be considered at the outset and in a proactive manner, the same way that genetic issues were with the *Universal Declaration on the Human Genome and Human Rights*, the *International Declaration on Human Genetic Data*, and the *Universal Declaration on Bioethics and Human Rights* at the infancy of genetic technologies, protecting privacy and working to prevent future harms.

I advance two ideas to encourage greater accuracy in the neuro-modification research so as to mitigate potential harm. First is a call for greater definitional specificity so that it is clear what one is and is not attempting to modify. What is meant by religion or moral enhancement? Whose agenda and definition of morality is driving moral enhancement? Second, I encourage greater recognition of the distributed and multifunctional nature of many brain processes, recognizing that cognitive localization is not the only story, and this involves, at minimum, that neural modification techniques be required to disclose potential side-effects in the same manner as medical drugs.

I have also highlighted several rights: the right to cognitive liberty, the right to mental integrity, and the right to psychological continuity. The articulation of these concepts in the law is not new. As they currently stand, however, I believe they permit a wide range of coercive and unconsented intrusions into a person's neural processes and correlated thoughts, emotions, and behaviors. These rights should be re-examined and reconceptualized, taking the new and emerging technologies of neuro-modification into account. Additionally, the current exceptions to rights dealing with the mental sphere should also be reassessed given their generality and broad possibility of interpretation. Finally, I advocate for an internal criterion that should assist individual researchers in developing helpful technologies with lesser potential for abuses: that of 'acceptability across ideologies'. The potential for harm, collateral risks, unintended negative consequences, and coercive and unconsented intrusion into people's mental domains is high with the emerging field of neuro-technology, so it is pertinent to prepare the field in advance.

References

Agar, N., 2010. Enhancing genetic virtue?. *Politics and the Life Sciences,* 29(1), pp. 73-75.

Bartels, A. and S. Zeki., 2004. The neural correlates of maternal and romantic love. *Neuroimage,* 21(3), pp. 1155-1166.

Beauchamp, T. and J. Childress., 1983. *Principles of biomedical ethics.* New York: Oxford University Press.

Beauregard, M. and V. Paquette., 2006. Neural correlates of a mystical experience in carmelite nuns. *Neuroscience Letters,* 405(3), pp. 186-190.

Bublitz, J. C., 2013. My mind is mine!? Cognitive liberty as a legal concept. In: E. Hildt and A.G. Franke, eds. *Cognitive Enhancement: An Interdisciplinary Perspective.* New York: Springer, pp. 233–264.

Davis, C. F., 1989. *The evidential force of religious experience.* Oxford: Oxford University Press.

Frank, M. J. et al., 2007. Hold your horses: impulsivity, deep brain stimulation, and medication in Parkinsonism. *Science,* 318(5854), pp. 1309-1312.

Freeman, W. J., 1995. *Societies of brains: A study in the neuroscience of love and hate.* Hillsdale, NJ: Lawrence Erlbaum Associates.

Gauthier, I. et al., 2000. Expertise for cars and birds recruits brain areas involved in face recognition. *Nature Neuroscience* 3(2), pp. 191-197.

Holbrook, C. et. al., 2016. Neuromodulation of group prejudice and religious belief. *Social Cognitive Affect Neuroscience,* 11(3), pp. 387-394.

Houeto, J. et. al., 2002. Behavioral disorders, Parkinson's disease and subthalamic stimulation. *Journal of Neurology, Neurosurgery, and Psychiatry* 72(6), pp. 701–707.

Ienca, M. and R. Andorno., 2017. Towards new human rights in the age of neuroscience and neurotechnology. *Life Sciences Society and Policy* 13(1), 27 pages.

Kapogiannis, D. et. al., 2009. Cognitive and neural foundations of religious belief. *Proceedings of the National Academy of Science USA* 106(12), pp. 4876-4881.

Klaming, L. and P. Haselager., 2013. Did my brain implant make me do it? Questions raised by dbs regarding psychological continuity, responsibility for action and mental competence. *Neuroethics* 6(3), pp. 527-539.

Mackenzie, R., 2011. Who should hold the remote for the new me? Cognitive, affective, and behavioral side effects of dbs and authentic choices over future personalities. *AJOB Neuroscience*, 2(1), pp. 18–20.

Newberg, A. et. al., 2001. The measurement of regional cerebral blood flow during the complex cognitive task of meditation: a preliminary spect study. *Psychiatry Research*, 106(2), pp. 113-122.

The Nuremberg Code., 1949. In: A. Mitscherlich and F. Mielke, *Doctors of infamy: the story of the Nazi medical crimes.* New York: Schuman, pp. xxiii-xxv.

Persinger, M. A., 1983. Religious and mystical experiences as artifacts of temporal lobe function: a general hypothesis. *Perceptual and Motor Skills*, 57(3), pp. 1255-1262.

Persson, I. and J. Savulescu., 2008. The perils of cognitive enhancement and the urgent imperative to enhance the moral character of humanity. *Journal of Applied Philosophy*, 25(3), pp. 162–177.

Pham, U., et. al., 2015. Personality changes after deep brain stimulation in Parkinson's disease. *Parkinsons Disease*, pp. 490-507.

Pycroft, L. et. al., 2016. Brainjacking: implant security issues in invasive neuromodulation. *World Neurosurgery*, 92, pp. 454-462.

Ramachandran, V. S. and S. Blakeslee., 1998. *Phantoms in the brain: Probing the mysteries of the human mind.* 1st edition. New York: William Morrow.

Singer, E., 2009. Targeting the brain with sound waves. *MIT Technology Review* [online]. Available at: <https://www.technologyreview.com/s/413697/targeting-the-brain-with-sound-waves/> [Accessed 27 June 2019].

Stein, B. E. and M.A. Meredith., 1993. *The merging of the senses.* Cognitive Neuroscience Series. Cambridge, MA: MIT Press.

Thompson, R. F., 1990. Neural mechanisms of classical conditioning in mammals. *Philosophical Transactions of the Royal Society of London B: Biological Sciences*, 329(253), pp.161-70.

Uttal, W.R., 2001. *The new phrenology: The limits of localizing cognitive processes in the brain.* Cambridge, MA: MIT Press.

Walker, M., 2009. Enhancing genetic virtue. *Politics and the Life Sciences*, 28(2), pp. 27-47.

Weissenbacher, A., 2016. The neuroscience of Wesleyan soteriology: The dynamic of both instantaneous and gradual change. *Zygon* 51(2), pp. 347-60.

White, L., 1967. The historical roots of our ecological crisis. *Science* 155(3767), pp. 1203-1207.

List of Abbreviated Terms

Arterial Spin Labeling (ASL)

Artificial General Interligence (AGI)

Artificial Intelligence (AI)

Artificail Intelligences (AIs)

Artificial Superintelligence (ASI)

Blood Oxygen Level Dependence (BOLD)

Brain Computer Interface (BCI)

Catholic News Service (CNS)

Cerebral Blood Flow (CBF)

Clustered Regularly Interspaced Short Palindromic Repeats (CRISPR)

Deep Brain Stimulation (DBS)

Default Mode Network (DMN)

Electroencephalography (EEG)

Erythropoietin (EPO)

Fédération Internationale de Football Association (FIFA)

Functional magnetic resonance imaging (fmri)

Genetic Virtue Project (GVP)

Human Level Machine Intelligence (HLMI)

Intelligence Augmentation (IA)

Latent Semantic Analysis (LSA)

Lysergic acid diethylamide LSD

Magnetoencephalography (MEG)

Major Depressive Disorder (MDD)

Major League Baseball (MLB)

Medial PFC (MPFC)

Medial Temporal Lobes (MTL)

MIDI (Musical Instrument Digital Interface)

National Health Service (NHS)

Natural Language Processing (NLP)
N,N-dimethyltryptamine (DMT)
Obsessive Compulsive Disorder (OCD)
Organisation for Economic Cooperations and Development (OECD)
Parahippocampal Cortex (PHC)
Parahippocampal Gyrus (PHG)
Pontifical Academy of Life (PAL)
Positron Emission Tomography (PET)
Posterior Cingulate Cortex (PCC)
Prefrontal Cortex (PFC)
Primary Visual Field (V1)
Quality of Life (qol)
Radio Frequency Identification (RFID)
Resting State Functional Connectivity (rsFC)
Retrosplenial Cortex (RSC)
Secondary Visual Field (V2)
Single Photon Emission Computer Tomography (SPECT)
Spiritually Augmenting Brain Computer Interface (SABCI)
Structural Equation Models (SEM)
Survival Risks (s-risks)
Task Performance Network (TPN)
Transcranial Magnetic Stimulation (TMS)
United States Conference of Catholic Bishops (USCCB)
Universal Basic Income (UBI)
Ventral Prefrontal Cortex (VPFC)
Visual Field 3 (V3)
Visual Field 4 (V4)
World Anti-Doping Agency (WADA)

About the Contributors

Christopher Benek is the Pastor and CEO of The CoCreators Network. He is a Presbyterian Church (U.S.A) Pastor who has served churches in New Jersey, Ohio, South Carolina and Florida. Across the globe, he is considered a leading expert regarding Artificial Intelligence and Emerging Technology. The Rev. Dr. Benek is regularly featured in media sources worldwide. He is a frequent speaker who is internationally known for his social and religious commentary. You can read more about him at www.christopherbenek.com.

Jacob Boss is a doctoral candidate and associate instructor in the Department of Religious Studies at Indiana University Bloomington, and editorial assistant at the Journal of American Academy of Religion. His research focuses on transhumanism and the do-it-yourself human augmentation movement.

Philip Reed-Butler is Visiting Assistant Professor in the Theological Studies Department at Loyola Marymount University in Los Angeles, California. His work primarily focuses on the intersection of neuroscience, technology, spirituality and race. His is the author of "Making Enhancement Equitable: A Racial Analysis of the Term 'Human Animal' and the Inclusion of Black Bodies in Human Enhancement" (*Journal of Posthuman Studies*, 2018). He recently completed his first book project entitled *Black Transhuman Liberation Theology*, which imagines what might happen if Black people utilized technological advancements to enhance both Black spiritualities and Black bodies in the struggle of materializing liberating realities.

Michael Caligiuri is an alumnus of the University of Manitoba and the University of Ottawa, where he earned degrees in both the sciences and humanities, and a PhD in Religious Studies. He is a Research Fellow of St. Paul's College and an instructor in Catholic Studies and Religious Studies at the University of Manitoba. His areas of focus include religious and secular bioethics, issues in body modification technology, cybernetics and nanotechnology, as well as science, ethics, and religious systems.

Irene J. Dabrowski is Associate Professor of Sociology at St. John's University, Staten Island Campus, New York City, where she served as Chair of the Division of Social Sciences for a decade. Dr. Dabrowski's main areas of expertise are futurology, the sociology of health and illness, urban sociology, women's studies, and the sociology of education. Her research and teaching incorporate interdisciplinary, holistic, and systems thinking. She has published an article (with Anthony L. Haynor), "Valuing the Future," in a World Future Society volume. As a Visiting Scholar at The Hastings Center (a think tank that focuses on biomedical ethics), Dr. Dabrowski investigated the ethical dimensions of holistic health care, integrating the work of the quantum physicist David Bohm into this project. Since 2005, she has served (with Anthony L. Haynor) as Co-Coordinator of the New Jersey Chapter of the World Future Society. Dr. Dabrowski is an advisory board member of the Lifeboat Foundation, an international think tank that assists humanity in addressing the existential risks and misuses associated with technology. As a member of SENCER (Science Education for New Civic Engagements and Responsibilities), she has attended summer institutes on science education and its implications for student civic engagement. Dr. Dabrowski is as a Liaison Board Member of the East Coast Colleges Social Science Association (ECCSSA) and has served as Associate Editor of *The ECCSSA Journal,* where she published an article on the post-positivist paradigm that called for a revision of social science in response to the Singularity. Her recent scholarship has critiqued artificial intelligence and a transhumanist worldview in an emerging electronic civilization, drawing on the insights of Catholic Social Thought.

Mark Graves is Visiting Research Assistant Professor at University of Notre Dame's Center for Theology, Science, and Human Flourishing with his research occurring at the intersection of artificial intelligence, psychology, and theology. He earned his doctorate in computer science at the University of Michigan in the area of artificial intelligence, completed a postdoctoral fellowship in genomics at Baylor College of Medicine as one of the first computer scientists to work on the Human Genome Project, and worked in biotechnology and pharmaceutical research for ten years before studying systematic and philosophical theology at Graduate Theological Union (GTU) and Jesuit School of Theology at Berkeley. He has published forty technical and scholarly works in computer science, biology, psychology, and theology, including the books *Designing XML Databases* (2002), *Mind, Brain, and the Elusive Soul* (2008) *and Insight to Heal: Co-Creating Beauty Amidst Human Suffering* (2013) and taught courses engaging the relationship between science and religion at Santa Clara University, the Graduate Theological Union, University of California Berkeley, Fuller Theological Seminary, and the University of Notre Dame. His current research in cultural analytics and machine ethics uses semantic text analysis to create computational models of human morality.

Anthony L. Haynor is Associate Professor of Sociology in the Department of Sociology, Anthropology, and Social Work at Seton Hall University in South Orange, New Jersey. He served for several years as department Chair. Dr. Haynor's main areas of interest are sociological theory, social problems, self and society, the sociology of knowledge, and social change. In his book, *Social*

Practice: Philosophy and Method (Kendall/Hunt, 2003), Dr. Haynor presented a communitarian method for social problem-solving. More recently, he contributed a chapter, "Classical Sociological Theory," to the *Cambridge Handbook of Sociology* (2017). Dr. Haynor has published (with Irene J. Dabrowski) in the area of futurology and has given presentations at the World Future Society Annual Meetings. He has also published (with Irene J. Dabrowski) on Catholic Social Thought, most recently on the contribution of the document, *Gaudium et Spes*, to an analysis and assessment of the transhumanist movement. Dr. Haynor is currently working on integrating the human sciences in the service of maximal human flourishing in a civilizational epoch characterized by ever increasing instrumental rationality, demographic diversity, concentrations of wealth and authority, and global interdependence.

Christopher Hrynkow, PhD (Peace and Conflict Studies, University of Manitoba), ThD (Christian Ethics, University of Toronto) is an associate professor in Religion and Culture at St. Thomas More College, University of Saskatchewan. From January 2019 to July 2019, he is serving as Patrick and Barbara Keenan Visiting Chair in Religious Education at the University of St. Michael's College in the University of Toronto. In July 2019, he will return to St. Thomas More College as Department Head and Undergraduate Chair in Religion and Culture for the University of Saskatchewan. There, Hrynkow also teaches courses in Religious Studies, Catholic Studies, Peace Studies, and in the Critical Perspectives on Social Justice and the Common Good program.

Ray Kurzweil is one of the world's leading inventors, thinkers, and futurists, with a thirty-year track record of accurate predictions. Called 'the restless genius' by *The Wall Street Journal* and 'the ultimate thinking machine' by *Forbes* magazine, he was selected as one of the top entrepreneurs by *Inc.* Magazine, which described him as 'the rightful heir to Thomas Edison'. PBS selected him as one of the 'sixteen revolutionaries who made America'.

Ray was the principal inventor of the first CCD flat-bed scanner, the first omni-font optical character recognition, the first print-to-speech reading machine for the blind, the first text-to-speech synthesizer, the first music synthesizer capable of recreating the grand piano and other orchestral instruments, and the first commercially marketed large-vocabulary speech recognition. Among Ray's many honors, he received a Grammy Award for outstanding achievements in music technology; he is the recipient of the *National Medal of Technology*, was inducted into the *National Inventors Hall of Fame*, holds twenty-one honorary Doctorates, and honors from three U.S. presidents. Ray has written five national best-selling books, including New York Times best sellers *The Singularity Is Near* (2005) and *How To Create A Mind* (2012).

Braden Molhoek works at the Center for Theology and the Natural Sciences, is a Lecturer in Science, Technology, and Ethics at the Graduate Theological Union, and is a Lecturer in the School of Engineering at Santa Clara University. He teaches courses on science and religion, ethics, software ethics, and bioethics. Having been interested in the intersection of science and religion since he was a double major in genetics and religion, Molhoek has published and presented on a variety of topics, including genetic engineering and virtue, theological anthropology in light of science, cloning, gene patents, and transhumanism.

Peter Robinson is Professor of Computer Technology at the University of Cambridge, where he works on problems at the boundary between people and computers. This involves investigating new technologies to enhance communication between computers and their users, and new applications to exploit these technologies. His recent work has included desk-size projected displays, emotionally intelligent interfaces and applications in semi-autonomous vehicles. This has led to broader explorations of what it means to be human in an age of increasingly human-like machines.

Una Stroda is a musician and a theologian. A native of Latvia, she currently resides in Chicago, IL. She holds degrees in piano performance, in ecumenical studies, in cross-cultural theology, and is a recent PhD graduate from the Lutheran School of Theology at Chicago. Her dissertation explores biblical, historical, and theological aspects of laughter in relation to the divine: the laughing God, the presence and absence of laughter in scriptures and the Christian tradition, and eschatological perspectives of human laughter.

Tracy J. Trothen is a professor of ethics at Queen's University, jointly appointed to the School of Religion and the School of Rehabilitation Therapy where she teaches in the graduate Aging and Health Program. She is an ordained minister in The United Church of Canada, a certified Supervisor-Educator in Clinical Spiritual Health (CASC), and a Registered Psychotherapist (CRPO). Trothen's areas of research and teaching specialization include: embodiment, biomedical and social ethics, Christian theology, spiritual health, aging, human enhancement technologies, and sport. Trothen is the author of *Spirituality, Sport, and Doping: More than Just a Game* (2018). Her other recent books include *Winning the Race? Religion, Hope, and Reshaping the Sport Enhancement Debate* (2015), and the anthology *Religion and Human Enhancement: Death, Values, and Morality* co-edited with Calvin Mercer (2017). She is currently at work, with Calvin Mercer, on a study guide tentatively entitled *Living Healthy for 500 Years and Other Technological Enhancements: Heaven or Hell?* Trothen is a member of the American Academy of Religion's Human Enhancement and Transhumanism Unit Steering Committee.

Alan Weissenbacher served many years as a counselor to homeless addicts, removing them from the urban setting and empowering them to run a farm while receiving counseling, spiritual care, and job training. His work with these clients inspires his research into neuroscience and spiritual formation, exploring ways to improve religious care and addiction recovery through understanding how the brain works. Recent publications include the chapter on neuroscience and religion in the textbook, Religious and Science edited by Gary Ferngren and an article exploring the neural correlates of instantaneous and gradual religious change published in *Zygon*. He is the book review editor for *Theology and Science* and is a father to two young boys.

Index

A

B

C

D

E

F

G

H

Printed in the USA
CPSIA information can be obtained
at www.ICGtesting.com
CBHW071356191124
17646CB00019B/248/J

9 781622 738236